LATIN AMERICAN STUDIES

Social Sciences and Law

Edited by
David Mares
University of California,
San Diego

A ROUTLEDGE SERIES

LATIN AMERICAN STUDIES: SOCIAL SCIENCES AND LAW
DAVID MARES, *General Editor*

OBSERVING OUR *HERMANOS DE ARMAS*
U.S. Military Attachés in Guatemala,
Cuba, and Bolivia, 1950-1964
Robert O. Kirkland

LAND PRIVATIZATION IN MEXICO

Urbanization, Formation of Regions, and Globalization in *Ejidos*

María Teresa Vázquez Castillo

Routledge
Taylor & Francis Group

LONDON AND NEW YORK

Published in 2004 by
Routledge
2 Park Square, Milton Park,
Abingdon, Oxon OX14 4RN

605 Third Avenue,
New York, NY 10017

Routledge is an imprint of the Taylor & Francis Group, an informa business

First issued in paperback 2012

Library of Congress Cataloging-in-Publication Data

Vázquez Castillo, María Teresa.
    Land privatization in Mexico: urbanization, formation of regions, and globalization in *ejidos* / by María Teresa Vázquez Castillo.
        p. cm.—(Latin American studies. Social sciences and law)
    Includes bibliographical references and index.

    1. Ejidos—Mexico.   2. Ejidos—Law and legislation—Mexico.
3. Land tenure—Government policy—Mexico.   4. Privatization—Mexico.
I. Title.   II. Series: Latin American studies (Routledge (Firm)). Social Sciences and law.

    HD1289.M6V39   2004
    333.3'172—dc22   2004001249

ISBN 13: 978-0-415-65535-4 (pbk)
ISBN 13: 978-0-415-94654-4 (hbk)

A mis papás, a mis hermanas y a
mis sobrinas y sobrinos, por permitirme
la ausencia de estos años.
A mi abuelita, una mujer Indígena
en la Ciudad de México.
A los Zapatistas, por su lucha por la
tierra y el derecho a un espacio.

Contents

Contents

List of Tables

List of Tables

List of Figures

List of Maps

List of Maps

Acknowledgments

I wish to acknowledge the communities, colleagues, friends, and institutions that, in different stages of my research, gave me their support to design, carry out, and complete this research. Profound gratitude goes to the *ejido* communities of Ixtaltepec, La Poza, and San Luis Río Colorado that sheltered me, assisted me, and taught me during my stay in their communities. I also want to thank Alfonso Guzmán and Humberto Aburto, planners of the city of Acapulco; Stephen Joseph, planner of the state of Sonora; the staff of the newspaper *El Sur*; and the staff of PROCEDE in Oaxaca and Sonora, all of whom provided invaluable data and insights. Special thanks to Chema Gamboa and Nemesio Rodríguez, in Sonora and Oaxaca, respectively. At Cornell, I want to thank my dissertation committee that underwent special health circumstances and survived. I want to recognize Susan Christopherson's effort to heal herself, which meant that she could read and keep on directing this research. I want to express my gratitude to María Cook, who shared her friends and colleagues in Mexico with me and whose special support has always been present. I thank Max Pfeffer and Barbara Lynch for being Lourdes Beneria's proxies during the process of this research. This research was funded with grants from the InterAmerican Foundation, the *Ejido* Project of the University of California at San Diego, the City and Regional Planning Department at Cornell University, and the Beatrice Brown Award of the Women Studies Department at Cornell University. In different stages of this research I was kindly welcomed and housed at the Latin American Studies Program at Cornell University, the Center for Mexican American Studies of the University of Texas at Arlington, and the Department of Ethnic and Women Studies at the California State Polytechnic University. My thanks go to Debbie Castillo, Mary Jo Duddley, Manuel García, and Patty de Freitas, directors of those institutions for their support. My gratitude also goes to friends who housed me, fed me, and took care of me during field research, especially to Daniel Coss Rangel and Amalia, in Sonora, to Margarita Ordaz in Guerrero, and to Gudrun Dohrman and Ms. Musalem in Oaxaca. I want to thank all my friends who read initial, intermediate, and final drafts of proposals and manuscripts, especially Joan Ormondroyd, Myrtle Bell, and Roberto Treviño. Joan became my second mother and unconditional editor in Ithaca; without her support this book would not be in your hands. Myrtle not only read my dissertation chapters, but she also took care of me in Texas. Roberto kindly agreed to edit part of my research. I also want to thank my friend, Greg Thrush, who helped me with some of the graphs and my many formatting questions. My appreciation goes to Ann Peters, Clyde Peña, Barbara Lynch, Takeko Inuma, Pilar Parra and Max Pfeffer for their friendship and for sheltering me while in Ithaca; to Davydd Greenwood for his unconditional support; to Phil McMichael for the food for thought; and to Betty Deakin for providing support in times when I needed it most. I also want to thank those friends who, even though they did not read any piece of this work, were there for me and supported me in the midst of a hostile but rich academic environment, especially to Agustín Andrade, Angélica Guerrero, Ann Forsyth, Armando Mejía, Betsy Sweet, Bob Marra, Brett Troyan, Christian Zlolniski, Debbie Castillo, Heli, Gilbert Cadena, Hiromi, Josephine Mahinda, Kate Ambrey, Leni Dharmawan, Loanna Valencia, Margarita Ledezma, Marina Trejo, Mary Jo Duddley, Noah Najarian, Rebecca Hovey, Reina Ramírez, Roger Frahm, Rosa López, Sanae Toyoda, Marina Trejo, Tony Humber, Verónica and Chris Kribs Zaleta, and to those who I do not mention but have a special place in my memory because their encouragement and support made this book possible. Finally, I thank the loving support of my family who encouraged me from Mexico City, especially my unique sister/friend Soledad whose love and encouragement were key in finishing this work.

Chapter One

Deregulating the *Ejido*: The New Article 27

In the 1990s, widespread deregulation and privatization were the processes that pervaded most economies in the so-called Third World, and Mexico was not an exception. Deregulation and privatization became the messianic recipes that would resolve the inadequacies of underdevelopment and its fiscal hurdles. Supposedly, these policies would bring democracy and efficiency to the economies adopting them. Deregulation and privatization became the policies that supranational institutions exerted as part of their economic restructuring programs to be applied around the world. Within this context, in 1992, the Reforms to Article 27, or *Ejido* Reforms, were some of the many economic restructuring policies carried out in Mexico. These reforms would deregulate *ejido* lands, permit their privatization, and open them for investment.

The original Article 27 was the constitutional legislation that regulated land and its redistribution from the end of the 1910 Revolution until 1992, when the Reforms were approved. These Reforms amounted to a counter-agrarian reform and meant for many a further wave of land privatization in Mexico. Such a dramatic reversal triggered a rainfall of questions about the fate of lands in Mexico, the impact of the new legislation on the well-being of the communities inhabiting those lands, and the implications of the Reforms at the local, urban, regional, and global levels. Those inquiries about the *ejido*, the meaning of its transformation, its deregulation and privatization, and its impact on communities were the initial issues that prompted this research.

This chapter introduces the organization of the research developed to find answers to those initial questions. It discusses the elements of both the original and reformed Article 27 and the theoretical framework used to approach this study. It also presents the hypotheses that directed this research, summarizes the chapters that sustain the case studies included in this book, and explains the criteria used to select those cases. Finally, it presents an overview of findings that this research generated. I will start by explaining the significance of the deregulation of the *ejido*.

THE EJIDO: THE PROMISED LAND

The Mexican *ejido* is a land tenure system that resulted from the Mexican Revolution of 1910 in which more than one million Indigenous people died in their struggle for land.[1] When propounded in the 1917 Constitution, Article 27 promised land to the landless and restoration of land to the displaced. In the political culture of Mexico, the *ejido* became mythicized as a "revolutionary" entity through which Mexican Indigenous people and impoverished peasants would have access to the land promised to them by means of the abolition of *latifundios*[2] and the redistribution of the land that their elimination would release.

More than 75 years after it had been written into law, amendments to Article 27 were announced in 1992, as part of a bundle of macropolicies intended to encourage investment and to "modernize" the countryside. These reforms represented a radical transformation of one of the most significant social contracts of the Mexican Revolution, that of land redistribution through the *ejido* system. Simultaneously, the 1992 amendments restructured the relations between the State and different social actors related to the *ejido*. Immediately, because of the agrarian character of the 1910 Mexican Revolution, this dramatic legislative shift posed a significant impact on agriculture and rural communities. More significantly, however, the reversal of land legislation on *ejidos* raised pressing questions related to ownership, future land uses, the local and regional economies sustained by *ejido* ownership rights, and the labor markets attached to the economies of *ejidos*. Another question relates to the influence of the *Ejido* Reforms in accelerating urbanization and globalization processes that were already transforming the rural character of *ejidos* and shaping urban areas and national and global regions. These questions arose because the *Ejido* Reforms basically erased the premises of the 1917 Article 27 and, consequently, its origin and history. In the next section, I will describe the past and present premises of Article 27.

ARTICLE 27 IN 1917 AND IN 1992

The original Article 27 of the Constitution of 1917 regulated land and its redistribution from the end of the 1910 Revolution until 1992, when the Reforms were approved. That legislation had as its main objectives:

- to regulate land redistribution and ownership in Mexico,
- to establish national sovereignty over all land, water, and natural resources,
- to restore land to Indigenous communities,
- to redistribute land to dispossessed rural communities, and
- to expropriate large land holdings for redistribution.

This legislation had created a form of land tenure and redistribution called the *ejido*. The *ejido* was defined as the legal entity of the "social interest sector," whose jurisdiction lay in the hands of Mexican-born peasants. *Ejido* lands were inalienable, nontransferable, and non attachable. In addition, *ejidatarios*[3] were the only ones able to own *ejidos*, which could not be conveyed, leased or mortgaged, or used as collateral for loans.

In contrast, the 1992 amendments to Article 27 or *Ejido* Reforms specified that:

- the State is no longer obligated to continue Agrarian Reform,
- the State is no longer obligated to provide land,
- the Mexican government has no power to expropriate land,
- *ejidatarios* are given the option to buy their own *ejido*, to lease it, to transfer it, to use it as collateral for loans or to mortgage it, and that
- *ejidatarios* can form associations or joint ventures with commercial groups.

PROCEDE

Along with the new reforms, a program called PROCEDE was implemented. Through this program *ejidatarios* obtained land certificates and titles. PROCEDE stands for *Programa de Certificación de Derechos Ejidales y Titulación de Solares Urbanos* or Program of Certification of *Ejido* Land Rights and Titling of Urban Lots. It was going to be a voluntary, participative program that could only be carried out if residents of the *ejido* so decided by voting and undergoing a sequential process of three consensual *ejido* meetings.[4]

The 1992 *Ejido* Reforms and the PROCEDE program were, in effect, a counter-agrarian reform that brought a further wave of land privatization to Mexico. Extensive research began focusing on the consequences of the reforms on rural Mexico, on agriculture and, to a lesser degree, on urban areas and urban populations. In the next section, I briefly review the main schools that have studied the *Ejido* Reforms.

RURAL AND URBAN APPROACHES TO THE STUDY OF THE 1992 ARTICLE 27

After 1992, several studies mushroomed collaboratively both in Mexico and in the United States to analyze the transformation of Article 27. At that time, it was possible to identify two major schools of research on this land legislation. One of them was what I have denominated the San Diego School, with David Myhre at its head. This school focused on the transformations of rural Mexico under the 1992 Article 27 and their impact on the Mexican rural economy.[5] The second school was what I call the Austin School, organized by Peter Ward. Ward et al approached the 1992 *Ejido* Reforms from an urban land perspective.[6] Little communication, probably because of their different approaches, existed between both schools. This might have been the reason that no further linkage developed in terms of the rural-urban interaction of the *ejido* transformation.[7]

Even in Mexico, extensive rural and agricultural studies related to *ejido* land existed. However, the urban domain had been analyzed to a lesser degree. The communal character of the *ejido* had always been a question of debate among the Mexican economic and political elites.[8] Since the *ejido*'s institutionalization after the 1910 Revolution, this debate prompted studies, theses, reports, and other writings arguing for and against the *ejido*. Most of these studies, however, have dealt with the rural and agricultural aspects of the *ejido*. The emphasis on the rural character of *ejidos* has a simple explanation; most of them were rural and devoted to primary economic activities. In addition, the cultural construction of the *ejido* was closely intertwined with rural landscapes and populations; however, these were only some of the multiple representations that the *ejido* had for different populations.

THE MULTIPLE REPRESENTATIONS OF THE EJIDO

The *ejido* not only represented land, but also a historic and socio-economic space for different actors. Table 1.1 summarizes some of the main representations of the *ejido* for different stakeholders over the last 90 years. This table was created taking into consideration the different representations developed and explored in each chapter of this book, which is why each representation indicates the number of the chapter where it has been analyzed.

Thus, the *ejido*, its development, and its land, have taken on multiple representations in the past 90 years. Originally, the *ejido* was land that was "gained" through the 1910 Revolution. For several decades thereafter, the word *ejido* brought to mind the image of rural Mexico and its peasants. Officially and unofficially, the Revolution, rural Mexico, peasants, and Indigenous communities were key elements in the construction of the Mexican national identity. *Ejido* redistribution was a formal contract between the State and peasants (see Chapter Two). The *ejido* is also a community of people organized around the space that the *ejido* provides.

However, both the space and these people have changed over time, especially in those *ejidos* close to urban areas. Before 1992, the *ejido* provided an affordable "illegal" option for low-income communities to access land for housing (see Chapter Three). Over time, *ejidos* near urban areas became land open to invasion by urban squatter settlements. For this group, the *ejido* became the "seed of their cities." The distribution of urban *ejido* land to squatter settlers, under the form of "regularized" land, was an unexpected outcome of Agrarian Reform. For the State, the *ejido* represented the pool of land on which to build the projects that the State classified as pertaining to the "public interest." The *ejido* was also the State's political and voting base in the countryside. Before 1992, *ejidatarios* could not, by law, convert

TABLE 1.1   Multiple Representations of the *Ejido* in the Past 90 Years

Social Actor	Representation
Rural and Indigenous Communities	Land gained through the 1910 Revolution (Chapter Two) Access to the land from which they were previously dispossessed (Chapter Two) Community of people organized around the space that the *ejido* provides (Chapter Two)
Rural Elites and Caciques	Space of political control and authoritarianism (Chapters Six and Seven)
Urban Populations	Space of urbanization (Chapter Three) *Ejidos* close to urban areas have provided an affordable "illegal" option for low-income communities to access land for housing (Chapter Three) *Ejidos* near urban areas became land open to invasion of urban squatter settlements (Chapter Three) Seed of cities for squatters (Chapter Three) The distribution of urban *ejido* land to squatter settlers, under the form of "regularized" land, was an unexpected outcome of Agrarian Reform (Chapter Three)
Land Speculators	"Captive" land (Chapters Two & Three) Before 1992 the *ejido* could be converted into a commodity only by illegal means (Chapter Two) After 1992 the *ejido* could be converted into a legal commodity (Chapter Two)
State	Symbol used in the construction of the Mexican identity (Chapter Three) The *ejido* land redistribution was a contract between the State and peasants (Chapter Two) Pool of land on which to build the projects that the State classified as pertaining to the "public interest" (Chapter Three) State's political and voting base in the countryside (Chapters Two & Three) State's political and voting base in squatter settlements (Chapter Three)
Mexican People	Image of rural Mexico, peasants and Indigenous communities (Chapter Two)
National and Global Capital	Recipient space for national and global capital (Chapter Four)

Source: Elaborated by author.

their land into a legal or sellable commodity; the *ejido*, therefore, represented "captive land" in the eyes of national and international land speculators. However unintended, the *ejido* functioned as a growth control mechanism. Within the NAFTA[9] framework, the 1992 *Ejido* Reforms provided the legal basis upon which to commoditize *ejido* land by permitting its conversion to a legal, private possession. Thus, the *ejido* has become a recipient space for national and global capital (see Chapter Four), creating in this way new spaces of investment. These, then, are some of the many representations of the *ejido* considered when selecting the *ejidos* and when drafting the hypotheses for this research.

THEORETICAL FRAMEWORK:
REGIONAL AND URBAN PERSPECTIVE

The multiple representations of the *ejido* and the many actors who had some relation to its land, history, and location led me to conceptualize a framework that included those representations, the social actors involved, and the historic relevance of the *ejido* for the development of cities and national and global regions. In addition, having been in contact with both schools of research on *ejidos* in Mexico and in the United States,[10] along with my planning background, made me reconsider the 1992 Article 27 through an urban and regional perspective. This perspective has as its point of departure the fact that land provides the vital physical space where we live and work. It then considers land as the site of housing, labor, and economic activity. This perspective proposes to go beyond the rural and agricultural arenas and contextualize Article 27 within different spheres of analysis, these being the local, the national, and the global.

To consider further uses of land beyond the agricultural or rural sphere, we are led to conceptualize the *ejido* as space for urbanization and for the location of secondary and tertiary activities. Land conversion from rural to urban uses indicates urbanization processes. In the same vein, land conversion from rural to manufacturing or other economic uses is the element that points to the creation of economic regions. Based on these tenets, it is evident that the 1992 Article 27 is defining the current urban and regional trends in Mexico and, consequently, the pattern of reproduction of its capitalism.

Thus, the framework used to study the meaning of the Reforms in their multiple representations examines changing land use in Mexico **from the perspective of its interaction with the local, the urban, the regional, and the global contexts.** This framework is supported by two complementary theoretical approaches. The first one is Brian Roberts'[11] approach in his study on regions in Mexico and the second one is the framework that Edward Soja[12] developed about the importance of space.

According to Roberts, regions in Mexico are shaped **by their internal and external relations** that, in turn, have shaped land tenure, the organization of their labor force, and the arrangements of their urban centers.[13] Based on this approach, I conceptualize that not only regions, but cities and *ejidos* too, are shaped by their internal and external relations. Thus, my point of departure was to consider regions, cities, and *ejidos* as interconnected spaces both in their physical aspect as well as in their economic, political, and social relations. Then my framework proposes to conceive the land where the region, the city, and the *ejido* are settled not only in physical terms, but in terms of the relations established with the different actors inside and outside that land.

Another theoretical influence comes from Soja who–based on his reading of Lefevre[14]–proposes "that understanding the world implies, in the most general sense,"[15] an "awareness of the simultaneity and interwoven complexity of the social, the historical, and the spatial, their inseparability and interdependence."[16] He then argues that "the local and the particular are becoming simultaneously global and generalizable."[17] Based on Soja's approach, I believe that in order to understand the meaning of the *Ejido* Reforms, we must start by understanding the history of land in Mexico, the social relations that different actors have established around land, and the importance of the location and geography of that land. Thus, my framework proposes to conceive land use in Mexico in terms of the economic restructuring it has undergone historically and in terms of the social relations that such restructuring has generated. What I want to do by using Soja's approach is to expose how the Reforms to Article 27 produce a recomposition of economic regions, a restructuring of social relations, and a redefinition of the geography of land and space both inside and outside Mexico. My purpose is to show that the *Ejido* Reforms are part of a global process of privatization of the land and space of different rural and Indigenous communities which results in the production of the space of investment for global capital. The elements of this theoretical framework contributed to the formulation of the hypotheses of this research.

HYPOTHESES

The four main hypotheses to be tested by this research are:

1. My central hypothesis is that the changes in land policy represented by the 1992 Article 27 have deeper consequences in *ejidos* than just those of "modernizing the agricultural sector." I suggest that the *Ejido* Reforms have **profound urban, regional, and global consequences**.
2. Article 27 formalizes, ratifies, and accelerates a long-standing process of *ejido* land privatization. Consequently, the Reforms to the *ejido* law will increase the degree and pace of *ejido* land **privatization**, accelerating existing processes of **urbanization** and the **recomposition of the regions** to which these *ejidos* belong. Due to the uneven development of regions in Mexico, the transformations prompted by the *Ejido* Reforms will affect unequally the well being of *ejidos*, **exacerbating the current regional differences in Mexico**.
3. *Ejido* agricultural land will continue being transformed into other uses. *Ejido* **land conversion** implies an internal redefinition of the local economies, the labor attached to them, its power hierarchies,

and its internal organization, including gender relations. Depending on how the *ejido* has been inserted in the Mexican and regional economies, different trajectories in the evolution of these internal relations will be evident.

4. The recent wave of privatization of *ejido* land generated by the 1992 Article 27—the privatization of the *ejido* space—contributes to the **creation of new spaces for the location of national and global investment.** The different hierarchies in which national and global State institutions influence the privatization of *ejido* land and the relocation of global investment inform the recomposition of the nation State and the emergence of supranational regulatorial State institutions.

THE CASE STUDIES

The case study method was selected to approach this research because of the existence of a wide variety of *ejidos*, not only geographically, but also economically and ethnically. It was a way to try to make sense of the *Ejido* Reforms and their impact, taking into account that at the beginning of the application of the titling program 28,058 *ejidos* and agrarian communities[18] existed all over Mexico. I attempted to choose *ejidos* that would be representative case studies of the changes provoked by the application of Article 27 and the PROCEDE program.

My aim was to develop three case studies that could describe and explain what was occurring at the moment the 1992 *Ejido* Reforms were in the process of being implemented. These case studies followed the approach outlined at the beginning of this chapter. That approach consisted in making the global, regional, urban, and local connections in *ejido* transformation.

Since my research questions focused on whether and how the application of the 1992 Article 27 was changing the *ejidos* in Mexico. I needed to first find out if changes were taking place, how they were taking place, and the reasons that provoked those changes. Based on these questions I selected the case study approach because it would allow me to trace change in *ejidos* through field research and by interviewing residents in the *ejido*. Or as Yin puts it, ". . .in brief, the case study allows an investigation to retain the holistic and meaningful characteristics of real-life events. . ."[19]

The *ejidos* selected belong to specific regions, are devoted to specific economic activities, have a specific relation to the PROCEDE program, and all of them show signs of transformation. Thus, they could typify similar cases of transformation of *ejidos* in terms of land privatization, urbanization, economic restructuring, and the creation of new investment spaces for national and global capitals at the expense of the destruction of the local economies. In Stake's terms, with the *ejido* as my unit of analysis, I tried to "generate knowledge of the particular."[20]

My case studies were designed to address and place historically each of the *ejidos* selected. This research had the objective of giving voice to members of the *ejido* by incorporating their history and describing the transformation of their communities.

In addition to the quantitative data collected from the census and the data generated from the questionnaire applied in the *ejido*,[21] qualitative material was gathered in the form of participant observation, ethnographic notes, and life stories of older members of the *ejido*. These stories contributed to putting together the history of the *ejidos* and to tracing their transformations from the perspective of their residents. How change in these *ejidos* has come about could have not been learned by exclusively reviewing statistical and archival data on land conversion or population change. Rather, it required ethnography and participant observation. I realized that, in order to assess the current reality of the *ejido* in its internal and external relations, it was necessary to utilize a range of research methods that could lead to identifying those relations.[22] The multiplicity of methods used in this research is indicative of my belief that the transformation of the *ejido* was caused by multiple elements. The objective of the fieldwork was to look within each *ejido* for the multiple causes of its transformation.

My field research took place between 1995 and 1996, in the midst of the certificating and titling program (PROCEDE). The approach employed to study each of the three *ejidos* consisted in doing background research followed by relevant interviews in the city closest to where the *ejido* was located, to examine conditions in some other *ejidos* in close proximity for comparison, and finally to focus research on the selected *ejidos*. In the nearby city, I gathered statistical and census data about the *ejido*, reviewed newspapers and *ejido* documents, carried out unstructured interviews with key informants from outside the *ejido*. Among those outside informants were planners and public officials from the PROCEDE program and other related institutions such as the *Registro Agrario Nacional* (National Agrarian Registry), the *Secretaría de la Reforma Agraria* (Ministry of Agrarian Reform), and CORETT.[23]

Within the *ejido*, I first interviewed *ejido* authorities. Although I had a list of the main topics to ask about the *ejido*, the interviews with the *ejido* authorities were open-ended and informal. I also interviewed the residents of the *ejido* and carried out participant observation. I designed a structured interview questionnaire that focused on the internal economy of the *ejido*, conversion of the land use of the *ejido*, and on changes in activities that *ejido* residents were experiencing. My objectives were to answer questions related to the privatization and transformation of the *ejido* from within.

I used a snowball sample to identify about 20 households for interviews per *ejido*.[24] The purpose of the structured interview was to determine the impact of the 1992 Article 27 on each selected *ejido*. To this effect, the questionnaire asked *ejido* populations whether they knew about PROCEDE.

Other questions targeted the status of their internal and external economic and social relations in order to get a sense of how the *ejido* was changing. The questionnaire was based on talks with Carlota Botey, a Mexican scholar and policy-maker with ample experience on Mexican Agrarian Reform and rural development. I had originally planned to review the roster of *ejidatarios* in each *ejido* and to select randomly 20 households. However, due to the unreliability of *ejido* records, I eventually decided to go home by home and undertake the interviews in those households that were available.

THE SELECTION OF EJIDOS

The field research had the objective of determining the ground level impact of the 1992 Article 27 on *ejidos* and their communities and to document changes, if any. The criteria established to select the three case studies where I carried out fieldwork assumed that *ejidos* reflect regional differences and, for that reason, this research designed a regional comparative study of three different *ejidos* in three different regions in the Western part of Mexico: one *ejido* in the North, a second one in the Center, and the last one in the South.[25] The purpose of this selection was to learn how regional differences were expressed in the character and degree of the impact of the 1992 Article 27 Reforms. At the time I initiated field research, in 1994, the application of the 1992 Article 27 was incipient; therefore, I focused my selection on those *ejidos* that were already undergoing rapid transformations in land and labor. These were *ejidos* that historically had been located closer to cities or that belonged to regions with leading economic activities.

In each of the selected states, Oaxaca, Sonora, and Guerrero, I traveled directly to the major city and visited its periphery. In the state of Guerrero I went to Acapulco, which is larger than its capital city of Chilpancingo, and I also visited the nearby *ejidos*.[26] In all three states, the booming urbanization of the *ejidos* close to the city was evident. However, the *ejidos* surrounding the tourist city of Acapulco were experiencing the most dramatic and violent changes of all. The changes here fit my selection criteria of rapid transformation of land use.

In Hermosillo, Sonora, I interviewed planners and public officials in charge of carrying out different stages of PROCEDE.[27] They recommended in turn, visits to several agricultural and coastal *ejidos* that were transforming their lands for commercial fishing and tourist purposes. During my stay in Hermosillo, I witnessed several protests by people living in nearby *ejidos* who wanted that land to be "regularized" without their being displaced. These urban protests were indicative of the pressure on *ejido* land in the periphery of Hermosillo. However, my interviews and visits to different Sonoran *ejidos* helped me determine that it was neither the peripheral nor the agricultural and coastal *ejidos* that were the ones undergoing the most rapid change. Actually, in Sonora, the border *ejidos* led the most rapid transformation.[28]

To find a rapid-transformation *ejido* in Oaxaca was not an easy task as most land is ordered under the agrarian community land tenure system that Indigenous communities have preserved over time. Thus many of the communities I visited were not, it turned out, *ejidos*, but agrarian communities.[29] Around the city of Oaxaca, both *ejidos* and agrarian communities were the source of land for urbanization in its periphery.[30] Nonetheless, activists, planners, and personnel of PROCEDE recommended that I visit the *ejidos* in the isthmian oil producing region as they were growing and urbanizing most rapidly. The fastest growing cities in that area were Salina Cruz and Juchitán. I selected the area around Juchitán as I had established previous connections there and it fell within the suggested rapid-growth region.

Once I had identified the three kinds of *ejidos* (coastal, border, isthmian) to be studied in each state based on their rapid conversion to other uses, it was necessary to narrow down the selection to a single *ejido* in that state. The deciding element was their degree of participation and reaction to PROCEDE; the measuring, subdividing, and titling program through which the 1992 Article 27 was to be implemented.[31]

San Luis Río Colorado was the very first *ejido* to participate in the PROCEDE program to officially privatize its lands. On the opposite end, the *ejido* of La Poza had refused to participate in the PROCEDE program as they saw it as fiscally damaging for the future of the *ejido* residents. Finally, PROCEDE personnel had already measured the lands of the Ixtaltepec *ejido*, which was the first stage of the titling program. In Ixtaltepec, PROCEDE had started the measuring process, at that time, without the knowledge and consent of the Indigenous population of the area (see Table 1.2).

Thus, the three *ejidos* selected were: the border *ejido* of San Luis Río Colorado in the Northwestern state of Sonora; the coastal *ejido* of La Poza in the Centralwestern state of Guerrero; and the Isthmian *ejido* of Ixtaltepec in the Southwestern state of Oaxaca (see Map 1.1). Interestingly, the *ejidos* selected also covered the most important sectors of the Mexican economy: the *maquiladora* industry (in San Luis Río Colorado), tourism (in La Poza), and oil extraction and subsistence agriculture (in Ixtaltepec).

TABLE 1.2 The Three *Ejidos* Selected

Ejido	*San Luis Río Colorado (Sonora)*	*La Poza (Guerrero)*	*Ixtaltepec (Oaxaca)*
Region	Northwestern	Centralwestern	Southwestern
Geography	Border	Coastal	Isthmian
Economic Base	Maquiladora	Tourism	Subsistence agriculture and oil
Participation in PROCEDE	Privatized its lands	Refused to participate in the program	The program started without the consent of the *ejido*

Source: Based on exploratory field research.

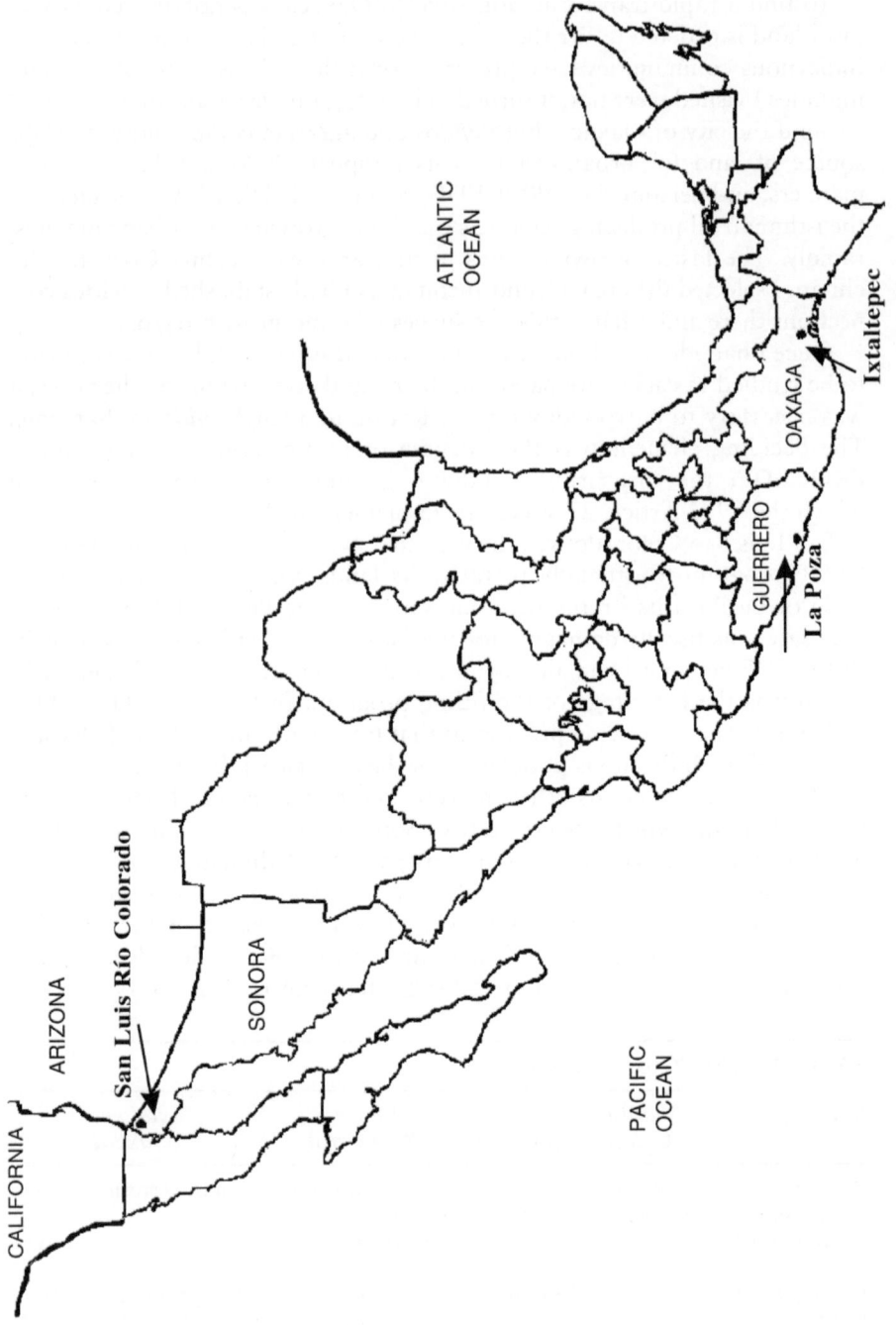

Map 1.1 Three *Ejidos* in Three Different Regions

CHANGING THE CONTRACT: CLASS, ETHNICITY, AND REGIONS

Based on the theoretical framework, my hypotheses, and the *ejidos* selected, the organization of this research followed a sequence of history, land policy, social relations, and regional relevance at the local, urban, national, and global levels. Within this context, the objective of Chapter Two, *Origins and Transformation of Article 27*, is to disprove the argument that the 1992 *Ejido* Reforms will not generate land privatization. It does so by contextualizing Article 27 and the *ejido* system within the history of land tenure and land policies in Mexico. It is important to trace ownership and dispossession over time in order to explain that privatization is not a new process, but rather a concomitant action occurring at each stage of the history of land in Mexico. Framed in those terms, the 1992 *Ejido* Reforms are just another land policy intended to further privatize Indigenous and rural lands in Mexico. Until 1992, *Ejido* lands had been protected by legislation that prohibited their sale. The 1992 Reforms to Article 27 wiped out the principles of land redistribution to the landless and the return of lands to their Indigenous owners. Throughout history, privatization has signified land deprivation for rural and Indigenous communities and land appropriation for the elite in power.

Another objective of Chapter Two is to elucidate the struggles and dynamics among different ethnic groups, classes, and regions in Mexico, always keeping in mind that the colonial past has colored these struggles until today. Thus, this chapter emphasizes that the actors participating in the 1910 Revolution represented classes, ethnic groups, and regions that have shaped the current geography of Mexico. Along with the agency of those groups, a combination of economic and land policies has affected the formation of regions in Mexico, defining an uneven development where the North and the South correspond to the rich and the poor, respectively.

THE *EJIDO*-CITY-REGION INTERACTION

In order to sustain this particular argument, it is first necessary to understand the role that land has historically played in Mexico as well as in its relation to the formation of cities and regions. History is essential to frame the land policies that the Mexican State has imposed through time on the different actors connected to land in Mexico. On the one hand, tracing Mexico's colonial past allows us to explain the authoritarian land use planning that continues today. On the other hand, it also points to the different tenure relations that affected the land and its inhabitants. As Chapter Two traces the colonial history of the *ejido*, Chapter Three on the *Urban and Regional Dimensions of Article 27* explores the historical trajectory of *ejidos*, cities, and regions as interconnected spaces that recreate an *ejido*-city-region-city-*ejido* interaction. This interaction considers *ejidos*, cities, and regions as spaces that mutually transform as internal or external relations among them shift.[32]

Chapter Three provides the framework to consider the *ejido* not solely as a rural entity but rather as a place that, due to formal or informal land privatizations, historically has been pushed into urbanization and industrialization processes. This chapter explains that the *ejido* is a fundamental component of the processes of urbanization in Mexico inasmuch as Mexican cities grew and expanded on its land. In the same vein, the *ejido* gradually converted from their agricultural uses in order to house increasingly differentiated economic activities. Therefore, the *ejido* is closely linked to the formation of cities and regions in Mexico and, consequently, the 1992 Article 27 influences the processes of urbanization and regionalization in Mexico. It instigates change by accelerating the conversion of rural *ejidos* into other uses and by simultaneously altering the local and regional economies sustained on that land. The 1992 land legislation represents a new but yet familiar stage of rural-urban transformation in Mexico. It represents a restructuring of space that affects not only the local and national, but also the global geographies. Chapter Four, entitled *Privatization of Ejido Land in the Age of NAFTA*, explains the global aspect of this legislation.

THE ROLE OF SUPRANATIONAL INSTITUTIONS

Following the analysis in Chapter Three that highlights the interaction between the internal and external economic relations of the *ejido*, the city, and the region, Chapter Four, *Privatization of Ejido Land in the Age of NAFTA*, passes to the next level, the global. It does so by scrutinizing the mechanism through which different supranational institutions influenced the implementation of this land macropolicy in Mexico. The role of the World Bank, the International Monetary Fund, and private international banks was key in binding future lending to the fulfillment of the economic restructuring program. The Reforms to Article 27 were part of that program whose main elements were privatization and clarification of property rights. The measuring and titling program of the 1992 Article 27 clearly accomplished those goals.

This chapter proposes that Article 27 was reformed in order to offer cheap land, natural resources, and labor to the global capital that searched for new markets and new spaces of investment. The Mexican State carried out privatizations and modified legislation so that investment could move freely and settle smoothly. In fact, the real meaning of free trade was the elimination of barriers to free investment. The application of the *Ejido* Reforms in the age of NAFTA disguised the history of dispossession of Indigenous communities in Mexico as well as the reasons that originally created Article 27. The formulation and implementation of the 1992 Article 27 diminished the importance of the national State and swelled the role of supranational institutions in the formulation of policies in Mexico. In fact, the Mexican State reorganized its functions in order to lay the background for further land privatization.

The working of the local, national, and global circumstances is deeply implicated in the understanding of the current transformation of *ejidos*. While Chapters Two through Four provide the history, social relations, and role of national and global circumstances in order to situate the field research accomplished in Mexico after the approval of the *Ejido* Reforms; Chapters Five through Seven present each case study in its internal and external economic relations.

THE *EJIDOS*

Chapter Five explores the case of San Luis Río Colorado in the Northern state of Sonora. San Luis Río Colorado was the first *ejido* that privatized land under the auspices of the 1992 Article 27 and its PROCEDE program. It was also the first *ejido* that constituted a so-called *ejido* real estate partnership with foreign capital which made it appear as the successful example of what *ejidatarios* could achieve by privatizing their lands. In reality, this *ejido* was not fulfilling the pretended modernization of the agriculture sector about which the supporters of the Reforms to Article 27 boasted during the approval of its amendments. Rather, the investment this *ejido* was attracting was converting its agricultural lands into a mega industrial park with real estate development for the future establishment of additional *maquiladoras*, an industry that mushroomed in the Northern region of Mexico due to its strategic proximity to the United States.

La Poza, Guerrero, is the case study examined in Chapter Six. La Poza used to be an *ejido* devoted to agriculture and fishing. The *ejidatarios* have also tried coconut oil extraction and lately, due to land conversion, plant nurseries. The development of the tourist industry in the Acapulco region has violently displaced and relocated *ejido* populations. The State has carried out these displacements of populations by expropriating *ejido* land in the name of the "public utility" and against the will of *ejidatarios*; this policy has generated extreme violence in the state officially considered the poorest in Mexico. La Poza refused to fully participate in PROCEDE and just wanted to receive their *ejido* certificates but not their titles for their urban lots, which are the lots their housing occupies. Historically, coastal *ejido* land has been expropriated for tourism purposes and subsequent reorganizations of land and restructuring of the local economies have taken place. PROCEDE would not contribute to the well-being of *ejidatarios* in this region as it was the State itself that colluded with national and foreign capital to expropriate the land these investors required.

The last *ejido*, examined in Chapter Seven, was Ixtaltepec, Oaxaca, located in the Southwestern portion of the Tehuantepec Isthmus, near Juchitán City. Although this was a predominantly agricultural and Indigenous *ejido*, the oil industry developed in the region attracted labor from nearby *ejidos*, including Ixtaltepec. Indirectly, the oil industry generated an emigration of population from the *ejido* Ixtaltepec as well as privatization of *ejido* land. PROCEDE started in this *ejido* without the required consent and knowledge of all of its *ejidatarios*. This lack of knowledge about policies and decision making affecting *ejidos* is a constant in all *ejidos*, but mostly in those where the Indigenous presence is stronger.

OVERVIEW OF FINDINGS

My field research indicated that formal and informal privatization and land use conversion had taken place in the three *ejidos* selected, even before the approval of the 1992 *Ejido* Reforms. The modified Article 27 had a differential impact in each *ejido*. This impact can be explained by the way each *ejido* had developed historically and because of their location and resources.

Regional location, geography, regional economic base and investment, and the composition of their population were important elements in determining the land uses within the *ejido*. For example, the border *ejido* of San Luis Río Colorado in Northwestern Mexico had received capital investment for the location of the *maquiladora* industry. The location of this industry accelerated urbanization in the *ejido*, causing a process of "savage urbanization" that transformed the *ejido* into one of the currently fastest growing border cities. The *ejido*-city-region interaction was even more evident as San Luis became the first *ejido* to privatize land under the 1992 Article 27, which prompted further land use conversion, further urbanization, and definition of the *maquiladora* industry in the border region. In this *ejido*, the Reforms had an immediate impact on the restructuring of land and contributed to the reorganization of labor.

In the same vein, the Central West coastal region, where the *ejido* of La Poza is located, has received several waves of tourist investment that have gradually pushed native residents out of that precious coastal land. Privatization in this *ejido* has mainly taken place through violent expropriations directed by the State for the benefit of private capital. From the three cases examined in this research, the *ejido* La Poza has been the one undergoing an almost permanent reorganization of its land and labor as displacement of native populations continues today. Although the *ejido* population had refused to completely participate in the titling program of the 1992 Article 27, privatization through violent means persists.

In contrast, the Ixtaltepec *ejido*, located in the Southwestern part of Mexico, had received different waves of investment that influenced the economy of the region. The latest one had concentrated on the oil industry that since the 1940s has attracted labor from nearby towns and cities, including Ixtaltepec. This *ejido* was a clear example of the interaction and mutual transformation in the *ejido*-city-region relation. Although this *ejido* was close to Juchitán City, no further change was observed. However, urbanization maintained a steady pace. Since the application of the *Ejido* Reforms, foreign capital was flirting with local *ejidatarios* to induce them

TABLE 1.3   Three *Ejidos* in Three Different Regions (Overview)

	San Luis Río Colorado, Sonora	La Poza, Guerrero	Ixtaltepec, Oaxaca
Region	Northwest	Centralwest	Southwest
Geography	Border	Coastal	Isthmian
Regional Economic Base	Maquiladora	Tourism	Oil, Agriculture
PROCEDE	Privatized its lands	Refused to participate in the program	Program started without consent of the *ejido*
Land Privatization	Illegal sale for housing	Illegal sale for housing	Illegal sale for housing
	Expropriation for housing and manufacturing	Expropriation for tourism	
Labor Before			
1. Male	Agriculture	Agriculture & fishing	Agriculture & pottery
2. Female	Agriculture	Plant nurseries	Informal trade, food
Labor After			
1. Male	Agriculture & services	Services, tourism industry	Agriculture & oil industry
2. Female	Maquiladora industry	House cleaners, maids	Informal trade, pottery
Impact on Land	Land for industrial park Conversion for housing and services	Land for tourist resort, Vidafe Conversion in other *ejidos*	No major change, yet

Source: Data from the author.

to participate in a project to generate wind energy. This project required the privatization of *ejido* lands. Table 1.3 summarizes this overview of findings which Chapters Five through Seven will explain in more detail.

REORGANIZATION OF LAND AND LABOR

The transformation of land and the reorganization of labor are taking place as *ejido* land privatizes either formally or informally. As land converts into other uses different from the traditional agriculture or other primary economic activities, the labor attached to those activities is forced to change in order to survive. But labor is people; men and women. Under the paradigm of privatization this human factor seems to be completely disregarded.

Having in place the historical trajectory of *ejidos*, the struggles of different groups for access to land, and the different land and investment policies that shaped regions and cities, it should be highlighted that field research in *ejidos* was not complementary but rather it was key to understanding the consequences of Article 27. The effects of this legislation go from the global to the personal as it reorganizes different geographies and social contracts at the global, regional, urban, local, and personal levels. Some prefer to call this personal level the household. During field research, I found that different access to land resources shaped the gender division of labor in *ejidos*. Field research revealed another internal relation in the *ejido*. Within our framework of analysis this internal economic relation also contributes to the shaping of regions, cities, and the *ejido* itself. It also shapes and reshapes in different ways the lives of the populations affected by privatization and economic restructuring. Other scholars have begun to explore and analyze this sphere.[33]

RECAPITULATION

The objective of this research has been to understand the effects of the 1992 Article 27 on the well-being of populations. This enterprise has caused me to delve into different spheres of analysis and different types of approaches. History, social relations, and geography have been key to understanding how Article 27 emerged and changed through time. Gradually, this research told the history of land in Mexico and the struggles of different subordinated and elite groups in their quest for land. These groups represent not only different ethnicities and cultures, but also regional and economic interests.

Land conversion, urbanization, and the formation of regions in Mexico were processes that constantly appeared during my search for understanding Article 27 and *ejidos*. Once completed, my research revealed the history of Indigenous populations in Mexico and the profound racism that shaped the colonial context that has remained in force through the institutions that have regulated land. The history of Indigenous populations is not limited exclusively to rural areas as, when displaced from their lands, they migrated to urban areas. Thus, this research also pointed out the continuing struggle for land that reemerged once displaced populations migrated to urban areas.

The colonial past of Mexico seems always to be present in the land policies that have segregated the *Mestizo*,[34] privileged North, and the Indigenous, repressed South. Within this framework and in spite of the global forces that are pushing for a different result, the struggle for land by Indigenous populations has been the fight to exist in the current geography of Mexico.

When I initiated this research, discouraging comments asserted that it was too early to detect any effect of the law. Nonetheless, one of my objectives was precisely to identify those early effects as they would tell the history of the process of reorganization of capital, land, and labor. I found that land privatization in Mexico is not a new process. Privatization always entailed a "reorganization" of the State and its planning functions. What has changed through time are the roles that actors play, the institutions that carry out privatization, and the functions of the State in establishing the legal framework in which those actors and institutions relate.

Summarizing the process and results of this research, it is fair to assert that the 1992 Article 27 inaugurates a new stage of capital accumulation in Mexico as a new displacement of the rural labor force and its subsequent proletarianization are taking place. In consequence, a new stage of urbanization and regional reconfiguration is under way during this NAFTA era.

In this vein, the process of privatization of the San Luis Río Colorado *ejido* illuminates the processes of urbanization and integration in the border region with the United States where *maquiladoras* are mushrooming and prompting further urbanization. The case of the *ejido* La Poza elucidates the role of the tourism industry in coastal *ejidos* in prompting displacement and "savage" urbanization. Finally, the Ixtaltepec *ejido* sheds light on processes of urbanization in Southern Indigenous *ejidos* near growth poles such as the oil-producing regions.

In short, this research tells the history of *ejidos* and their populations at a specific point in time, that is, during the application of the 1992 Article 27. It is also the history of formation of new geographies and landscapes in Mexico in times of renewed globalization.

As the transformation in each *ejido* is related to its land, its labor, and the restructuring of its activities, I suggest that these cases explain the current transformation of regions, cities, and localities in similar parts of the world regardless of the organization of their land systems. Processes of privatization, land conversion, urbanization, and transformation of local economies and landscapes are taking place around the world as national and transnational investments find spaces to relocate.

Chapter Two

Origins and Transformation of Article 27 in Mexico

Article 27, the law that regulates land and natural resources in Mexico, is not an isolated piece of agrarian legislation that was suddenly modified in 1992. Rather, Article 27 reflects the struggles throughout Mexican history among different ethnic groups, classes, cultures, and regions. It contains within it the history of the colonized and the colonizer in their relation to land, space, and territory. Article 27 and the *ejido* system are the historical products of different phases of land privatization and of colonial and authoritarian land use planning that supported those privatizations. Thus, it is impossible to fully understand the significance of the transformation of Article 27 unless one also understands the history of land tenure in Mexico, and that of its landless and displaced communities.

The objective of this chapter is to provide the historical background for an analysis of contemporary processes of land dispossession and privatization and the influence they have had in the formation of Mexico's regions. In order to demonstrate that land privatization is not a new process, but rather a policy mechanism which has existed since the time of Spanish colonization, this chapter describes the emergence of the colonial *ejido*, as well as the colonial land policies that dispossessed Indigenous communities and privatized their lands. Ethnic colonial relations are explored as this chapter discusses the emergence and development of land struggles in Mexican regions, during and previous to the 1910 Revolution. This revolution lasted for more than ten years and caused the death of more than one million people, mostly Indians, who died in their struggle for land.[1] It also culminated in the writing of the Constitution of 1917.

Revolutionary policy makers included in Article 27 of that Constitution a reformulated notion of *ejido* that would simultaneously be the land unit and the tenure system through which land redistribution would be carried out. The *ejido* system and the past and present premises of Article 27 are developed in this chapter to illuminate their historical importance and their current significance under conditions of radical transformation.

Through the explanation of land tenure and agricultural policies, the chapter initiates a discussion about land transformations and regional differences in land use and labor markets. It argues that those policies, and the subsequent regional differences that they created, have shaped the development trajectories in Mexico, which in turn have created a privileged North and a disadvantaged South.

To describe the formation of these national regions, we must analyze the role of the State and its use of Article 27 to articulate a political and economic policy towards such different social actors in the countryside as Indigenous and landless peoples, *Mestizo*[2] populations, and Spanish-descent elites. The relation of the State towards these social actors was transformed in 1992 when Mexican Agrarian Reform—enacted through the application of Article 27 and the formation of *ejidos*—was revoked. This transformation denies the promise of access to land to the landless, and restoration of land to the displaced and legalizes the already existing privatization of *ejido* lands. These are the topics that this chapter develops.

THE COLONIAL *EJIDO* AND *BALDÍOS*

The *ejido* land tenure system existed prior to the 1910 Mexican Revolution under a different form. In early colonial times, Spaniards included *ejidos* (*exidos*) in their planning of cities and towns. Those *ejidos* were land grants for the use of city populations organized by Spanish colonizers. Later, this territorial organization also included Indigenous towns and cities. The colonial *ejidos* included, in addition to individual land grants for the members of a particular city or town, lands for common use or *propios* and *dehesas*. As time passed, the usage of the word *ejido* came to refer exclusively to those land grants for the common use of cities and towns.[3]

Although the Spanish Crown endorsed the right of Indians to their lands, at the same time it decreed itself as the holder of rights to all lands and waters in the colony, and treated unevenly the property rights of the Spanish and those of Indians. Spanish property could be sold, alienated or transmitted; Indigenous property could not. Furthermore, when the Crown decided that lands in Indigenous towns or cities were *baldíos* meaning "abandoned" or "not worked," it rescinded the Indigenous rights and granted the land to new owners, generally Spaniards. Those regulations, coupled with forced expropriations of Indigenous lands and the further displacement of Indigenous populations, resulted in an accumulation of large landholdings called *latifundios*.[4]

Crown public officials and the Church were the main benefactors in the accumulation of *latifundios* in colonial Mexico. By using their administrative or political positions, both groups illegally appropriated Indian lands.

Later, the Crown would legalize the ownership of those lands through a process called *composiciones*, which gave legal status of those lands and their property in exchange for payment to the Crown. When the *composiciones* payment was not made, the properties would pass to the Crown.[5]

By the beginning of 1810, just before the Independence Revolution, approximately 52 percent of total national land was considered *baldíos* or vacant lands. Most of these *baldíos* had once belonged to Indigenous communities that had been expelled or killed. That year, cities, *haciendas*,[6] and *ranchos* owned by Spaniards totaled 39 percent of the total Mexican territory, while only about 9 percent belonged to Indigenous communities.[7] In fact, this latter figure is high in comparison to the percentage of land that Indigenous communities ended up owning after the application of the 1823 and 1856 privatization laws of the post independence period.[8]

THE PRIVATIZATION OF INDIGENOUS LANDS

After the Independence Revolution of 1810 and within the context of the liberal ideas of the nineteenth century, private property emerged as one of the major premises of nascent capitalism. Mexican Liberal thought at the time emphasized the supremacy of legislation and of the State in regulating land ownership. During this period, internal struggles over land tenure and the size of land holdings were rampant among factions of Spanish-born landowners, Mexicans of Spanish origin, and the Church.

One outcome of those clashes was the passage of several privatization laws between 1823 and 1856, among them the Disentailment and Lerdo laws, which allowed the parceling and alienation of *baldíos* and promoted their colonization by foreigners. A second outcome was the parceling and privatization of the land owned by the Church and by Indigenous populations. Ibarra Mendívil explains that the objectives of those policies were to create a land market, to promote the sense of private property among Indigenous communities, and to reduce the political and economic power of the Church.[9] The laws, however, by giving foreigners privileges in the colonization of *baldíos*, had a clearly racial component in that they excluded Indigenous communities from access to lands that they had previously owned. In addition, the laws classified both the Church and the Indigenous Communities as "civil corporations" with no rights to legally possess land. Consequently, the laws affected, in the same degree, the lands of the Church and those of Indigenous communities, even though it was evident that the Church had concentrated enormous land holdings, while the Indians had been, since the Conquest, gradually dispossessed and expelled from their lands. Apparently, the land legislators simply ignored the disparate amounts of land that the Church and Indigenous communities owned at that time, and dealt equally with very unequal entities.

Land parceled and privatized through the Disentailment and Lerdo laws entered the market immediately and resumed a process of land concentration that had been taking place since earlier colonial times. For the Church this expropriation process was completely new, but Indigenous communities had learned, long before, the meaning of land privatization; the new laws applied during the Juárez regime just confirmed their knowledge.

In the Congress of 1857, some deputies opposed including Indigenous communities in the category of "civil corporations" claiming that those communities had the right to work and produce[10] and that access to their properties would allow them to exercise those rights. Although these deputies also supported private property, they argued that privatization would open lands to the market temporarily; however, after a certain time, lands would be reconcentrated in a few hands and kept unproductive. Their argument proved to be true.

When Tannenbaum tried to make sense of this period, he wrote:

> This theory of a recovery of the national lands was influenced by another belief. The idea of "progress" had been affected by a perverted social Darwinism, which held that in the struggle for survival only the fit survive, and the Indian was assumed to be unfit. The Indian was described in official publications of the Department of Agriculture as *un lastre*, a burden upon the economy of the land. It was believed that he ought to be displaced, his communities destroyed. The sooner the Indian and his ways disappeared from off the face of the land, the better; for with so great a burden upon it the country would not "progress" and become a modern state. A systematic campaign to destroy the Indian communities, and with them the Indian too, it was hoped, was initiated. The fact that at the time the Indian represented more than half of the total population was considered evidence of the need for the change rather than an argument against the enormity of the undertaking or the social hazards it involved.[11]

The post independence land privatization laws were reinforced by additional land laws prescribed under the dictatorial regime of Porfirio Díaz.[12]

THE SURVEYING COMPANIES AND THE NEW CONQUEST

A deep institutional racism against Indians, their culture, and their way of life was reflected in most of the land laws decreed before the Porfirio Díaz regime. However, it was under Diaz's mandate that the highest percentage of Indigenous lands were lost. It was also during his administration that an increasing number of Indigenous laborers lived in conditions of slavery so severe that they led to the extermination of such Indigenous communities as the Yaquis of Sonora.

The laws that enabled further dispossession of land from Indigenous communities and its monopolization by national and foreign big landowners were the 1883 Colonization Law, the Privatization of Vacant Lots (*baldíos*) and the Federal Water Laws of 1894. These laws stipulated that the colonization of *baldíos* could be carried out by the State or by private companies that were allowed to survey Mexican lands in order to privatize them and sell them.

It was some of Díaz's close allies, who belonged to the French-oriented Mexican elite of mostly Spanish origin, who obtained the concessions to form *compañías deslindadoras* or surveying companies. In turn, most of the members of that elite then transferred their concessions to foreigners, mostly from the United States.

Tannenbaum describes this process:

> . . . and it thus came about that a small number of foreign companies had a free rein to roam the country and examine the titles of all property-owners in Mexico. The companies were allowed to retain one third of all the "national" lands they discovered, and to purchase the rest at a pittance. It is true that they were required to colonize the lands they acquired, but, as a matter of record, they rarely did so. What in effect occurred was the transfer of a large part of the territory of the nation to a few companies, most of them foreign, many of them American. By the end of the Díaz regime 72,335,907 hectares (a hectare is equal to 2.47 acres), nearly a third of the Republic, had been surveyed. This was so great an upheaval in Mexican rural property that it has been called the New Conquest.[13]

Meanwhile some of the remaining two thirds of the land that surveying companies did not appropriate "were sold to wealthy landowners, foreign investors, and anyone who had money to pay for them."[14] Not satisfied with such preference, the Díaz regime exempted surveying companies and foreign investors from paying taxes. He applied a similar benefit to large landowners by not increasing their taxes.

Ibarra Mendívil characterizes the State under the Díaz dictatorship as "deeply interventionist" and privatizing in that it reorganized the economy for the benefit of large land owners and foreign and national investors. This reorganization required the destruction of Indigenous communal land ownership, which the State considered an obstacle to the "order and progress" of the modernization process being carried out by the Díaz regime. Ibarra Mendívil affirms: "The purpose of the intervention of the State was not to statize society, but rather to deepen its privatization."[15] Amazingly, that characterization accurately fits Mexico's current situation of economic restructuring and land tenure policies.

In the 1880s, land accumulation resumed through the surveying of lands, which promoted the ongoing privatization and transfer of Indigenous and national lands. Three groups benefited most from this transfer: the surveying companies, the individuals who transferred their concessions to foreign surveying companies, and big landowners who acquired some of the land on sale after the surveying process.[16]

By opening the door to U.S. ownership of Mexican land, the 1883 and 1894 land laws added a new foreign elite to the realm of big landowners in Mexico. Tannenbaum states that in some Northern states like Chihuahua and Sinaloa, foreigners, "many of them Americans," owned more than 40 percent of the total territory.[17] According to official records, by 1906,

50 surveying companies and big land owners held 47 million hectares, about a fourth of the national territory.[18]

When writing about conditions previous to the 1910 Revolution, Tannenbaum emphasizes that "[t]he Mexican Revolution cannot be understood without an insight into the irritation produced by the foreign ownership of land in the country."[19] Evidently, this "irritation was felt not only by dispossessed urban populations in pre-revolutionary Mexico, but also by Indigenous communities in rural Mexico as they experienced the loss of their remaining lands, additional disruption of their way of life, and their subsequent subjection to the *hacienda* system.

SLAVE INDIGENOUS LABOR AND CONCENTRATION OF LAND IN HACIENDAS

Haciendas were institutionalized estates that controlled both land and the labor force. During the Díaz regime they comprised about one half of the rural population and 82 percent of Mexican Indigenous communities.[20] By 1911, approximately "11,000 *haciendas* controlled 57 percent of the national territory, and 834 of these landowners held 1.3 million square kilometers in immense *haciendas*. At the same time, 15 million peasants, or 95 percent of rural families, were landless."[21]

As the privatization of Indigenous lands took place, the *hacienda* became the institution that absorbed the dispossessed labor force as *peones acasillados*. *Peones acasillados* were workers living in the *hacienda* on a permanent basis. They worked for the *hacienda* in exchange for some land to work: "one day's work each week during the year for the right to plant a hectare, two day's for two hectares."[22]

Because they were always in debt to the *hacienda* or the *hacienda* store, *peones acasillados* could rarely leave the *hacienda*. The *hacienda* owner obliged *acasillados* to buy from the *hacienda* store, charging them outrageously high prices. As a result, *acasillados* were forced to work for the *hacienda* in order to pay their debts which, no matter how much they worked, always increased and were transmitted from generation to generation. Because of their debts to the *hacienda*, their poor living conditions, their inadequate housing, their lack of rights and labor mobility, these displaced Indigenous laborers under the *hacienda* organization were the equivalent of slaves. *Haciendas* in Mexico were not unlike the slave plantations in the United States.

In summary, the following events and conditions contextualize and explain the 1910 Revolution: a history of conquest, dispossession, and displacement of Indigenous and other rural communities; an ongoing process of privatization and alienation of Indigenous lands for the benefit of the elite in turn; the manipulation of legislation to legitimize illegal practices of expulsion and ethnic cleansing; the supremacy and greediness of, first Spanish conquerors, and later of U.S. surveyors in their access to land and resources in Mexico; plantation and slave labor systems disguised under the names of *hacienda* and *peones acasillados*.

Thus, when the 1910 Revolution took place, "the emphasis was upon restitution."[23] Tannenbaum, when stating the objectives of Article 27 wrote:

> It must now be clear what the purpose of Article 27 was: To recover the land granted to large companies by concessions, to make it difficult for foreigners to acquire agricultural rural properties, to break up the *hacienda* system, to encourage the development of communities, to free the population from peonage, and to return to the villages the lands taken from them.[24]

REGIONAL, ETHNIC, AND CLASS STRUGGLES WITHIN THE REVOLUTION

Most conceptualizations of the 1910 Revolution considered that the majority of the armies participating in it were looking for land restitution and redistribution. This was partly true. Certainly the violent murders of revolutionary *caudillos* during the Revolution exemplified the struggles among them and the conflict among the political agendas they were advancing. These leaders were killed because they represented specific regional, ethnic and class interests, which were in constant opposition with each other when designing their demands from the Revolution.

For example, the opposition that Madero[25] initiated against the Díaz dictatorship comprised the urban middle class, some factions of the army, and industrial workers. His political agenda was to democratize Mexico by carrying out elections and by establishing an anti-reelection policy. However, the social environment in Mexico in the aftermath of the Porfirista regime, demanded changes not only related to the elimination of the dictatorial State, but more important, related to the privileged status that big landowners had enjoyed during the Díaz dictatorship as well as in earlier regimes. The pressing demand was agrarian in nature and was supported by different political and ethnic groups, namely indigenous communities and other Indigenous and *Mestizo* rural populations.

After the overthrow of Díaz, the agrarian policies that Madero pursued as president left the demands of rural populations unfulfilled. As a result, Zapata[26] emerged in armed opposition to the Madero regime.[27] At the same time that political support for Madero decreased because of his policies, army officials loyal to the Díaz dictatorship murdered him and installed Victoriano Huerta as his successor.[28] This prompted the appearance of the Carrancista[29] and the Villista[30] movements. Zapata, Villa and Carranza represented different political agendas and proposed radically different agrarian programs that, even today, reflect the uneven development of their regions of origin as well as the different profile of their populations. The trajectories of these three different movements provide a context for explaining the emergence of Article 27 of the 1917 Constitution.

THE ZAPATISTA MOVEMENT FOR AGRARIAN REFORM

The movement led by Emiliano Zapata emerged in the Central state of Morelos. In that region, sugar plantations had dispossessed Indigenous communities and other small land proprietors of their lands.[31] When Madero took over, those dispossessed communities expected him to issue policies to legally return their lands to them. That did not happen during Madero's tenure, so in 1911 they rebelled, declaring the *Plan de Ayala*. In this document they demanded the prompt restitution of lands to their original owners; that is, to Indigenous and other rural communities, and the redistribution of land for the landless. The demands advocated by the *Plan de Ayala* had widespread support throughout Mexico. Once the plan succeeded and the Morelos area recovered its lands, some of the *Zapatistas* refused to continue engaging in military operations outside Morelos. In general, however, the *Zapatista* movement continued fighting for agrarian change, even after the proclamation of Agrarian Reform stated in the 1917 Constitution. Zapata denounced the fact that, although Agrarian Reform had been adopted in the new Constitution, it was not being implemented. His continuing fight for agrarian change attracted crude military attacks from Carranza's armies and culminated in Zapata's assassination in 1920.[32]

THE VILLISTA MOVEMENT FOR SMALL PRIVATE PROPERTY

Francisco Villa represented a completely different region in Mexico, the Northern state of Chihuahua, where mining and cattle ranches predominated over agriculture. Chihuahua's population, like the rest of the Northern states, was composed mainly of Spaniards, *Criollos*,[33] or *Mestizos*. Ethnically, it diverged radically from the mostly Indigenous-populated South and Central *Zapatista* areas. The North had had a different historical trajectory. To illustrate the character of the region, Katz explains the role of military colonists in Chihuahua and their relation to land. These colonies were organized by the Spanish Crown in order to protect their Northern territories from the incursion of Apache Indians who had been constantly forced from their territory and way of life in the further North. Katz mentions that the Spanish Crown granted colonizers "extensive amounts of land; the right to freely buy and sell additional land; exemption from taxes; the right to administer their own towns and cities,"[34] and guns to stop the Apache Indians.

Around 1885, with the near extermination of the Indigenous group of Apaches and the construction of the railroad that connected Chihuahua to the United States and to Mexico City, U.S. "investors flocked into Chihuahua, buying both mines and large estates, and a new market for Mexican cattle emerged in the United States. Land values rose enormously, and for the first time the *hacendados* had a real incentive for appropriating for themselves so-called empty public lands and for expropriating the properties of the military colonists."[35]

With the application of the land laws under the Díaz dictatorship, the Northern native populations lost their grazing lands. Their political town and city organizations were also modified. In addition to surveying companies, "the richest family in Chihuahua, the Terrazas-Creel clan, enacted a new land law that forced the former military colonies to put up most of their lands for sale, which were then bought either by landowners or by wealthy members of these communities."[36] Although some revolts took place, they were defeated.

Foreign investment in the region developed mining and industrial factories. However, after 1907, the recession affected jobs in those industries. During the Revolution, military colonizers, unemployed workers and some sectors of the middle class, joined forces to support Villa and in 1913 he became governor of Chihuahua. Under his mandate, *haciendas* were placed under State control and their revenues were used to finance the revolution. His agrarian policy stated that he would distribute *haciendas* among his soldiers and return lands to towns. The most important difference between Villa and Zapata regarding agrarian reform was that Villa supported small private ownership of property, while for Zapata, redistribution respected the communal aspect of Indigenous lands, and the identity and interest of their constituencies. Another important difference was that Zapata effectively redistributed land of *haciendas*, while Villa kept most *haciendas* under his administration. Finally, Katz explains that because of the proximity of Chihuahua to the U.S. border, Villa secured income by selling cattle and agricultural products to the United States, which contributed to finance an army stronger than Zapata's.[37]

THE CARRANZA MOVEMENT: THE BIG LAND OWNERS

Venustiano Carranza took over in 1913, after the deposition of Victoriano Huerta, the army officer who had overthrown Madero. Carranza came from the Northern state of Coahuila whose history differed from that of the *Zapatista* Central and Southern regions or from the *Villista* state of Chihuahua. The *Criollos* and *Mestizos* of the Northern region—the states of Coahuila, Sonora, and Nuevo León included—had not been affected by massive usurpation of their lands. Rather, they had benefited from the violent displacement of such Indigenous communities as that of the Yaqui Indians in Sonora. Both Madero and Carranza were from Coahuila and belonged to *hacendado* families, which explains why agrarian demands were not at all their priority. If anything, they "wanted to maintain the structure of the *haciendas* and the *hacendados* as a social class."[38] Thus, Carranza's 1913 *Plan de Guadalupe* omitted any reference to land reform, nor did he redistribute or affect in any way the *haciendas*. His only "radical" measure was to increase taxes for foreign enterprises in the region under his control in order to generate revenues to support his military bases during the Revolution.

TABLE 2.1   Conflicting Political Movements in the Revolution by Region,
Ethnicity, Class, and Agrarian Agenda

Movement	*Region*	*Ethnicity*	*Class*	*Agrarian Agenda*
Zapatista	Central and South	Indigenous Groups and *Mestizos*	Small land owners and dispossessed rural populations	Agrarian reform and land redistribution
Villista	North (state of Chihuahua)	*Criollos* and *Mestizos*	Military colonists, miners, cattle ranchers	Agrarian reform and protection of small private property
Carrancista	North (state of Coahuila)	*Criollos* and *Mestizos*	Big land owners	No agrarian reform, no land redistribution

Note: Based on the historical information provided in Friedrich Katz, "The
Agrarian Policies and Ideas of the Revolutionary Mexican Factions Led by
Emiliano Zapata, Pancho Villa, and Venustiano Carranza" in Laura Randall, ed.,
Reforming Mexico's Agrarian Reform (Armonk, NY: M.E. Sharpe, 1996), 21–34.

In 1915, as he witnessed a weakening of his bases and the increasing
support for Zapata and Villa's movements, Carranza modified his agrarian
policy on paper by including the possibility of land expropriation and its
further redistribution. However, in practice, he constantly disengaged from
that policy, which prompted a strong response from the supporters of
agrarian reform. As a consequence, during the 1916 Constitutional Con-
vention, a majority of its members adopted Article 27.

The political circumstances of 1916–1917 forced Carranza to accept
the article on Agrarian Reform, but once he became president, he avoided
carrying it out. To the contrary, he amnestied émigré *hacendados*, returned
their expropriated land, and persecuted and eliminated *Zapatista* and
Villista supporters. On balance, agrarian reform occurred only in the areas
controlled by *Zapatista* and *Villista* forces. Table 2.1 summarizes these
three political agendas which shaped the 1917 Article 27.

ARTICLE 27 AND THE *EJIDOS* OF THE 1917 CONSTITUTION

The main objective of Article 27 of the 1917 Constitution was to regulate
land redistribution and ownership in Mexico and to establish national sovereignty over all land, water, and natural resources within its territory. The
ejido system was reintroduced in Article 27 as a means to expropriate large
land holdings, restore land to Indigenous communities, redistribute land to

dispossessed rural communities, and to achieve the goal of *Zapatistas*: to award "land for those who worked it."

Article 27 has historically been associated with communal lands; however, the Article protected and recognized three major forms of land tenure: private, public, and social. Evidently, Article 27 reflected the agendas of competing actors participating in the 1910 Revolution. Article 27 included communal and *ejido* lands as social property, while public lands comprised *baldíos* and other parceled and legally recorded land in favor of the State. In order to understand the meaning of Article 27, this chapter emphasizes the specific regulations established for the ownership of private and social lands.

PROTECTING PRIVATE PROPERTY

Article 27 made evident the conflicting agendas of the Zapatista movement for agrarian reform, the Villista movement for small private property, and the Carrancista agenda to preserve the properties of big landowners. On the one hand, Article 27 stated that private property could be expropriated and redistributed in order to disintegrate the *latifundios*; and, on the other hand, it also established legal protections for small holders of land. Consequently, private property, below a certain size, could not be expropriated. Article 27 defined private property as "land that did not exceed 100 hectares of irrigated land or the equivalent of 200 hectares of arid pasture land. In addition, one could own 150 hectares of land to cultivate cotton, 300 hectares to grow bananas, sisal, sugarcane, coffee, rubber, coconuts, grapes, vanilla or fruit trees or the amount of land necessary to maintain 500 head of cattle or other livestock."[39] In comparison to the amount of land allocated to individual *ejidos*, the legal size of private property was considerably larger.

In addition to allowing the expropriation of private property to be redistributed for the benefit of the communities requesting land, Article 27 also permitted the expropriation of private property in the name of the public good. However, through this research, we found that social property was first to be expropriated, while private property remained protected, except during the Cárdenas administration.

SOCIAL PROPERTY: COMMUNAL AND *EJIDO* LANDS

Communal and *ejido* lands were land tenure systems which existed prior to the 1910 Revolution. Communal lands were those lands exclusively adjudicated to Indigenous people, whose ownership was legally stated since times of the Spanish colony. *Ejidos* were the entities reestablished as the basis for Agrarian Reform in Mexico. Through *ejidos*, communities could organize in order to apply for land grants from the Mexican government. In contrast to the range of 100–500 hectares allocated to private property owners, the maximum extension of the individual *ejido* parcel could not exceed 10 hectares. Thus, land redistribution was weak from its start, since most of the solicitors of *ejido* and communal lands were of Indigenous origin and private proprietors were mostly *Criollos* and *Mestizos*. Access to land therefore continued preserving its ethnic and colonial character.

The *ejido* system granted land to communities rather than individuals;[40] as a result, the awarded land was owned collectively by the members of such a community. Over time, the word *ejido* became generalized and was used to designate both the population within the *ejido*, as well as the land itself and its tenure. In other words, the *ejido* was its people as well as their physical and social space; together they contributed to the identity formation of *ejido* populations.

The 1917 Constitution defined two types of *ejidos:* "The individual *ejido*, in which land tenure and ownership are legally vested in a community, but cropland is allocated on a semi-permanent basis among the individual *ejidatarios*, and the 'collective' *ejido*, in which land resources are pooled for collectively organized production. A majority of *ejidos* are of the individual kind."[41]

Mexico's colonial history and the interventionist and expansionist roles of foreign elites in Mexico caused the designers of Article 27 to define the *ejido* as a legal entity of the "social interest sector," whose jurisdiction lay in the hands of Mexican-born peasants.[42] "*Ejido* lands are inalienable, nontransferable and non attachable."[43] Supposedly, *ejidatarios* were the only ones able to own *ejidos*, which could not be conveyed, leased or mortgaged, or used as collateral for loans. However, crafters of Article 27 also included "exceptions" to this law. Once again, the conflicting agendas were present in establishing land ownership of the Mexican territory. Mexicans could own land, but the denomination of Mexican includes several ethnic, regional, and historical considerations. Who were the Mexicans owning land after the 1910 Revolution and who were the "exceptions?"

WHO COULD OWN LAND AND WHERE

The pernicious privilege that landlords from Spanish and later U.S. origin enjoyed in illegally appropriating land, coupled with the history of foreign armed interventions in Mexican territory, led the authors of Article 27 to declare the Mexican State as the owner of the land and natural resources, and establish that only natural-born, naturalized Mexicans, and Mexican companies had the right to acquire ownership of lands, waters and their accessions, or to obtain concessions of exploitation of mines and waters. Nonetheless, Article 27 also specified that the State was permitted to make exceptions and grant the same right to foreigners. In order to approve such exceptions, foreigners had to agree to be considered as nationals, before the Ministry of Foreign Relations and, consequently, to not invoke the protection of their governments for any matter relating to those properties. Otherwise, such proprietors would be penalized by having to forfeit in favor of the Mexican government the properties they had acquired.

The properties that foreigners could acquire through that exception clause could not be within a zone of 100 kilometers (about 62 miles) from the borderlands and/or 50 kilometers (about 31 miles) from the seacoasts, since those areas were defined as "prohibited or restricted zones."[44] These regulations became stricter during the Cárdenas administration (1934–1940) that passed a Law in 1934 excluding Mexican women married to foreigners and Mexican-born people whose parents were foreigners from acquiring land in the prohibited zone. The character of this law might have been the result of the tense process of nationalizing the oil industry that was mainly owned by U.S. capital, but it also reflects a matrilineal consideration regarding land tenure.

Undoubtedly, history was the main element considered by the crafters of Article 27 when specifying its different rules and regulations. They took into consideration the history of the continuous land conflicts among the Mexican government, the Mexican economic elites, and the Church.[45] The latter institution had for centuries enjoyed strong economic and political powers enabling it to concentrate large landholdings. In consequence, Article 27 emphasized the exclusion of religious groups from owning property and natural resources.[46]

In summary, Article 27 assigned original ownership of land, water, seas, natural resources, and sources of power and fuel to the State as representative of the nation. It also empowered the government to redistribute land, to expropriate private property in the public interest, and to decide on "exceptions" to the application of Article 27.

EXCEPTIONS UNDER ARTICLE 27: AVOIDING LAND REDISTRIBUTION

Although Article 27 conferred upon Mexicans and the Mexican State the right to own the national territory, the 1917 Constitution also included some exceptions to this statement. In the case of inheritance and judicial decree, for example, the Ministry of Foreign Relations was authorized to permit acquisitions based on pre-existing rights acquired in good faith; nonetheless, the property had to be conveyed to a qualified person within a period not exceeding 5 years. Over time and under certain conditions, foreign colonizers were also permitted to acquire land in the prohibited zone. It appears that these exceptions were the legal mechanisms implemented by Carranza and later presidents to return *hacienda* land that had been confiscated during the Revolution, invalidating in this way the redistribution that had taken place during combat. In a way, the exceptions reflect the gradual rise of the Northern elites and their ability to influence the regional trajectory of Mexico.

An additional legal recourse to avoid the application of Article 27 was the *amparo*. The *amparo*, which in Spanish means assistance, shelter or safeguarding, is a legal mechanism that could be filed for protection against State abuses. Private land owners, however, craftily used it to request exception or protection against expropriation of their lands. This *amparo* mechanism was introduced in 1949 and since then has been widely utilized by large landowners as a means to nullify the provisions of the Agrarian Reform as well as a way to keep ownership of their lands.

In 1966, a more recent exception was approved; the Ministry of Foreign Relations permitted the acquisition of prohibited-zone land by Mexican companies whose majority shares were owned by Mexicans. The 1973 Foreign Investment Law set a limit of 49 percent foreign ownership for any company operating in Mexico. This limitation was later revoked to encourage foreign investment. Thus, a foreign-controlled Mexican company might also be the principal shareholder in a company holding title to such land.[47] In addition, foreign capital or a foreign-owned Mexican company could lease prohibited-zone land for a term of ten years or less. This exception appears to have been drafted in order to legalize the establishment of the assembly line factories or *maquiladoras* on the border with the United States, as well as to legalize the nascent coastal tourist industry of that time.

Throughout the years, further exceptions have been added to allow foreign capital to acquire land outside the limits of the prohibited zone. The prerequisite was that "foreigners applying for such acquisitions had to be residents of Mexico, those holding tourist visas or transmigrant status could not be authorized to acquire property."[48]

On first reading, Article 27 appeared to be a very strict law regarding land tenure regulations; however, multiple exceptions have been gradually included in the legislation to facilitate investment in restricted areas. It is well known that foreigners continue to own sumptuous residences and huge land holdings in coastal areas, and that they continue to exploit a wide variety of natural resources such as mines, bodies of water, and forests. Meanwhile, many native populations continue to be dispossessed.

LAND PRIVATIZATION: EXPROPRIATIONS AND *PERMUTAS*

Perhaps one of the most important exceptions to Article 27 was the possibility of acquiring *ejidos* for private purposes. For this, Article 27 provided two mechanisms: expropriation and *permuta*, or exchange. The State could carry out land expropriations in the name of the public interest although its final destination was private. The *permuta* or exchange was an agreement between the State and *ejidatarios* consisting in awarding *ejidatarios* land in a different area in exchange for their original properties.

The advocates of such exceptions gave several reasons for expropriating communal land. One of the most relevant was the "creation, promotion, and preservation of an enterprise" in the interest of the community.[49] Those advocates comprised national and foreign capitals, which were gradually transforming Article 27.

It was evident that the land privatizing wing of the Mexican designers of Article 27 managed to gradually include a myriad of exceptions that, over time, were manipulated by unscrupulous lawyers, notaries, developers, and other officials to grant foreigners and foreign capital the right to acquire land in the prohibited zone as well as in other areas to the detriment of ownership rights of Indigenous communities. Evasion of Article 27 was common and facilitated by the fact that there were almost nonexistent enforcement mechanisms within the Mexican Legislation. The legal framework was in place, but planning, implementation, and enforcement mechanisms were weak. Rural Mexico witnessed anew the illegal acquisition of land and subsequent violent displacement of rural and Indigenous populations, and the impunity enjoyed by national and foreign *latifundistas* under different guises.

The exceptions to Article 27 were opposed by Indigenous and other rural communities who, for decades, denounced the Agency responsible for realizing the redistribution of land, the Ministry of the Agrarian Reform, for not performing its functions. They argued that those communities that had been lucky enough to be awarded land, had received lands that were useless for agricultural uses. As a consequence, the land had had no positive impact on their well-being. In addition, the agency's application of agricultural policies was inconsistent and favored a landscape conducive to deepening regional differentiation. Those policies provide the historical framework within which the 1992 *Ejido* Reforms were advanced.

AGRICULTURAL POLICIES AND REGIONAL DIFFERENTIATION AFTER THE REVOLUTION

Land redistribution in conjunction with agricultural policies implemented after the revolution, contributed to further define the Northern and Southern regions in Mexico. During the Lázaro Cárdenas administration (1934–1940), the distribution of land under Agrarian Reform reached its peak. Before and after Cárdenas, the unwillingness of different administrations to implement the commitments of Article 27 was evident. For example, the first postrevolutionary administrations of Plutarco Elías Calles (1920–1924) and Álvaro Obregón (1924–1928) were committed to the Northern agenda that protected big landowners. Thus, although one of the purposes of Agrarian Reform was to promote equity in rural Mexico, *latifundios* continued to exist and *caciques*[50] became a new caste of large landowners. Agrarian Reform hardly modified the overall economic structure in rural Mexico or the relations of power previously in place. *Latifundios* were still found everywhere in rural Mexico and inequalities in access to land remained. Initially, land redistribution was considered a revolutionary turn in the economic development of Mexico. However, for the land-privatizing wing of postrevolutionary administrations, Agrarian Reform was not considered a radical policy, but rather an emendation established to avoid further tensions with Indigenous communities and the rural sector.[51] Later administrations even perfected the mechanism of *agrarismo estadístico*, or statistical agrarianism through which approved presidential land grants were not awarded and in the few cases they were, grantees received land of poor quality.[52] Distributed or not, these lands appeared in the official statistics as though they had been granted.

The economic policies of the 1960s marked a shift from those of the 1920–1950 period. Land distribution slowed and the government established a wide range of agricultural subsidies and credit in certain regions. These allocations were designed to benefit those who produced agricultural products for export, primarily the Northwestern states of Mexico, such as Sonora and Sinaloa. They received water and electricity subsidies as well as credit, while in Southern Mexico *ejidos* continued producing as self-sufficient units using only their own resources. Public subsidies for irrigation projects increased the area of arable land mostly in the North. Producers in this region began utilizing fertilizers, pesticides, improved seeds, and mechanization in order to increase productivity and profitability. The process of regional differentiation continued with the 1970s agricultural policy, which favored the more capital-intensive agriculture in Northwest Mexico, in contrast to the labor-intensive practices of the South. As part of a policy to encourage exports, the government provided credit for wealthy *ejidatarios* in the North while many *ejidatarios* in the South were denied access to credit.

Within this context, when critics of the *ejido* system define it as a low productivity and inefficient form of land tenure and production, they forget to mention the policies that had induced "backwardness" and shaped its characteristics. Not all *ejidatarios* had access to resources and credit and, in many cases, the land they received was useless for cultivation. In addition, they have constantly been repressed by local authorities or by *caciques* of the region who regularly dispossessed them of their land or other resources.

During the mid-1970s, land distribution was restricted in the same way as it had barely occurred during the first postrevolutionary administrations. However, the government promoted new agricultural programs and policies. For instance, the Agricultural Development Law was passed to promote associations between *ejidatarios* and private landowners. In addition, the Mexican Food System (*Sistema Alimentario Mexicano* or SAM) was initiated as a policy to achieve self-sufficiency in grains. Government subsidies, combined with a propitious climate, produced favorable results by the beginning of the 1980s, but the oil crisis of 1981–1982 severely affected the SAM and subsidies decreased. During this period, agricultural policies and Agrarian Reform were also influenced by international institutions like the World Bank. As part of its restructuring policies the World Bank advised Mexico that, in order to receive funds, it had to "reduce at a gradual but drastic pace" subsidies to agriculture.[53] Consequently, the elimination of price guarantees and import licenses facilitated the entry of foreign products which began competing within the national market. At the same time, capital-intensive agriculture continued to emerge in the North and mechanization helped to displace rural workers in this region.

By the end of the 1980s we see the widespread presence of agribusiness in Mexico with production oriented primarily to the U.S. market—a process continued and speeded up by the policies carried out during the Carlos de Salinas de Gortari regime.

THE PRIVATIZATION PROGRAM OF THE 1990S

The central features of Salinas de Gortari's[54] land and agricultural policies were the elimination of State intervention in agriculture, the privatization of State agricultural businesses, and the approval of a new Article 27. This was the period in which the deregulation of the agricultural sector was completed. In spite of the withdrawal of most subsidies, a few continued in those regions favoring export-oriented agriculture. The Salinas plan to "modernize" Mexican agriculture offered support for those regions in the form of reduction of tariffs on agricultural inputs and credit. In fact, rural Mexico and agriculture were not defined as part of "the modern" economy. To even speak about the rural elements of the Mexican economy was equated, within neoliberal discourse, with backwardness. That philosophy did not change with the election of Ernesto Zedillo in 1994.

In fact, the Zedillo regime[55] continued the implementation of *Ejido* reform through the PROCEDE program,[56] initiated during the Salinas administration. Zedillo's objective was to complete the certification and titling of all *ejidos* and agrarian communities by the end of his administration. By March, 1998, two years before the end of his regime, the Annual Report of the *Procuraduría Agraria* reported that 60 percent out of a total of 27,144 *ejidos*[57] had been certified by the PROCEDE program.[58] A regional variation in the participation in the program was evident as more Northern and Central states than Southern states had completed the measuring and certificating process.[59] The willingness of *ejidatarios* to participate in the program and conflicts over land ownership and limits influenced that regional variation.

Although PROCAMPO,[60] a financial incentive program that supported regions and farmers, was put in place; it only benefited those "who decided to cultivate the most profitable commercial products."[61] The situation in the countryside did not improve during the Zedillo administration as he continued the policies of the Salinas administration. In the same vein, both the opposition organized by the *Zapatista* Army in the Southern state of Chiapas and the national debtors movement called *El Barzón*, that had appeared during the Salinas administration, also continued. The *Zapatista* movement of the 1990s exposed the authoritarian process by which Article 27 was approved and the negative effects of this policy on Indigenous and low-income rural populations. Concurrently, *El Barzón* organized significant opposition to unfair foreclosures of rural land by the banking system. Finally, during Zedillo's presidential period, one more guerrilla movement appeared, the *Ejército Popular Revolucionario* (EPR) or Revolutionary

Popular Army, also in Southern rural areas of Mexico,[62] bringing into question the modernization consequences of the *Ejido* Reform. So far, this policy has not been modified. The tenets of the new Article 27 and the rationale followed by policy makers to transform the Article underlie this questioning.

THE 1992 REFORMS TO THE *EJIDO* AND TO ARTICLE 27

In November 1991, the Salinas government proposed an initiative to amend Article 27 with the intention of "integrating" the Mexican economy into the global economy, "modernizing the Mexican agriculture," and "awarding adulthood" to *ejidatarios*.[63] The assumptions behind those objectives were that local and subsistence-oriented agriculture was considered backward and that *ejidatarios* had been treated as minors. On January 6, 1992, the federal government published this initiative and on February 26, 1992, the New Article 27 was passed by the Mexican Congress.[64]

The 1992 amendments to Article 27 specify that: a) the State is no longer obligated to continue with Agrarian Reform nor, consequently, to provide land; b) the Mexican government has no power to expropriate land to redistribute to peasants; c) *ejidatarios* are given the option to buy their own *ejido*, to lease it, to transfer it, to use it as collateral for loans or to mortgage it; and d) *ejidatarios* can form associations or joint ventures with commercial groups.[65] In short, these amendments deregulated the *ejido* land, placed it on the market, and opened it for investment.[66]

Along with the new reforms a program called PROCEDE was implemented. Through this program *ejidatarios* obtain land certificates and titles.[67] PROCEDE stands for *Programa de Certificación de Derechos Ejidales y Titulación de Solares Urbanos* or Program of Certification and Titling of *Ejido* Land and Urban Lots. It is a voluntary, participative program that can only be carried out if residents of the *ejido* so decide by voting and undergoing a sequential process of three consensual *ejido* meetings.[68]

In those meetings, *ejidatarios* would approve the intervention of the *Procuraduría Agraria* in their *ejido*. The *Procuraduría Agraria* is the administrative entity in charge of carrying out the PROCEDE program. This institution was created to measure and classify the parcels within the *ejido* as a prerequisite to awarding certificates and titles of *ejido* land. The measuring and titling processes were severely criticized by opponents of *Ejido* Reform.[69] They argued that as soon as *ejidatarios* obtained their certificates and/or titles they would be more prone to sell their land[70] in times of economic crisis (which are not uncommon in rural Mexico). They concluded that the changes to Article 27 made it easier for *ejido* land to be gradually converted into private property causing further displacement and dislocation of *ejido* populations. The Mexican State, however, insisted that the 1992 Reforms were implemented as part of a bundle of macropolicies intended to encourage investment[71] and integrate Mexico into the global economy.

RATIONALE FOR THE *EJIDO* REFORM

The "integration" of the Mexican economy into the global economy was the discourse used to justify the deregulation and privatization of the Mexican economy in general and of the *ejidos* in particular. The Mexican government has, for years, portrayed *ejidos* as inefficient units that needed foreign investment in order to be productive and vital.[72]

Further justification for this land macropolicy can be found in a recent document published by the Congress of the United States which characterized Mexican agriculture as inefficient and non-competitive.[73] According to information provided by this publication, most of the 4.5 million farms in rural Mexico are small and production consists mainly of corn and beans. In addition, peasants still cultivate using traditional techniques and most production is used for self-subsistence and local consumption. Only about 14 percent of Mexico's land is considered arable[74] and one third of that land is used for corn production.[75] Agriculture and Mexican *ejidos* are characterized as backward and non-diversified. However, the document did not explain the historical, technological, environmental or cultural elements of the *ejido*'s current state, nor did it explain that it was referring only to those *ejidos* whose economic basis was agriculture or agriculture-related. Neither did it point out the important fact that many of these *ejidos* are self-reliant.

Contrary to what the designers of the Reforms depicted about *ejidos* and their inefficiency, Barkin (1991) insists that after the Mexican Revolution, *ejidos* emerged as the producing units which could generate the most sustainable development for rural areas as well as self-sufficiency for communities and families. He describes how, during this period, Agrarian Reform provided credit, fertilizers, and technical assistance to small producers through the *Ejido* Credit Bank (*Banco de Crédito Ejidal*) and, as a result of this policy, Mexico became self-sufficient in corn by the end of the 1950s.[76] Therefore, the argument that *ejidos* are inefficient and backward is a generalization that does not take into consideration the history of *ejidos*, the region in which they were located, or the profound differences among them.

If "inefficiency" and "backwardness" are going to be the main arguments to eliminate Agrarian Reform, the designers of such a macropolicy first need to analyze how Agrarian Reform was applied. The cause of its "failure" was not the *ejidos* themselves nor their organization, but rather the historical vices that Agrarian Reform faced when trying to carry out the redistribution of land. For example, land applicants faced what Ibarra Mendívil characterizes as a "procedural labyrinth," a practice exercised by the bureaucracy of the Ministry of Agrarian Reform to control and delay the demands of land applicants. He states that complete application files were tabled, abandoned, or even lost.[77] Thus, procedures that should normally take between four months to a year, would often take a minimum of eight years. In some cases, they took between 13 to 20 years, only to receive a negative from that Ministry. Denouncers of these delays blamed the scarce personnel and the low budget assigned to the Ministry of the Agrarian Reform as causes of their inefficiency.[78] They pointed out that inefficient administrative procedures reflected the lack of willingness to carry out Agrarian Reform. Thus, *latifundios* were unaffected and they continued to exist in some areas, inequities in land distribution remained, and *caciques*[79] emerged as the new big landowners, both outside and within *ejidos*. Agrarian Reform did not modify the overall economic structure in rural Mexico nor change the power relations already in place. It seems, however, that more large land holdings will continue to emerge, since the new reforms permit the alienation of *ejidos* and many rural communities do not have the economic and political resources to keep them.

The final argument used to justify the *Ejido* Reforms was the possibility they provided to create conditions for *ejidos* to launch joint ventures with private capital. Barkin (1990) states "the joint venture that *ejidatarios* and Gamesa—a producer of cookies—launched in 1991, was highly publicized as an example of what the government expected from the implementation of the new agriculture policies." In this joint venture, *ejidatarios* leased their lands to private agribusiness. The new Article 27 states that associations with commercial or civil groups "unlike rural associations" can own *ejido* land.

Apparently, the criteria on which the Mexican government grounded the Reforms to Article 27 were exclusively related to rural *ejidos* and the agricultural sector. In the official documents that described the changes to Article 27, the Mexican government failed to describe the implications that this macropolicy would have beyond the rural sphere or the responses from those communities to the privatizing of land through the 1992 *Ejido* Reform. In later chapters the impact beyond the rural sphere is explored and analyzed, but the final section of this chapter discusses the responses of rural communities.

RESPONSES TO *EJIDO* REFORM
UNDER AUTHORITARIAN PLANNING

The way that State policies have been approved and implemented in Mexico reflects the lack of participation, of those who are directly affected, in the decision-making process. The character of the Mexican State has been, to a large extent, authoritarian, and, consequently so are its planning processes. The modifications to Article 27, for example, were proposed and approved behind closed doors. According to Fox, when the terms of Reform were made public, the Salinas administration "managed to persuade" peasant leaders to support them, either by suborning them or by threatening them.

Some the State indemnified as "lands officially ceded to claimants," but still in private hands; meanwhile, stronger opponents were openly threatened with exclusion from the already limited agricultural support programs.[80] Whatever the "choice" opted for by those leaders, there was a general lack of information about the Reforms among their constituents. So, when Fox emphasizes that "open protest of the Article 27 reform was minimal," he means that opposition was once again co-opted or repressed. In other words, the 1992 legislation for Article 27 was railroaded through with threats, bribes, and lack of information, behaviors typical of an authoritarian government.

The two "convincing" strategies exercised by the State illustrate the historical political relation developed between the State and the social actors within the *ejido*. Fox argues that since its reinsertion in the Constitution, the State has used the *ejido* simultaneously as: (1) an apparatus of political control; (2) a temporary political expedient and reserve migrant labor source; (3) a key pillar of a populist national development project; and (4) an organ of peasant representation.[81]

He asserts that rural populations in Mexico have been used for voting purposes in exchange for short-term benefits, a policy which has promoted a clientelistic relationship with the State. A case in point was the *Programa Nacional de Solidaridad* or National Solidarity Program (PRONASOL) implemented during the Salinas regime. Within this program, the municipal authorities "choose the recipients" of loans for public works.

In addition to the clientelistic relation created between the State and actors within and outside the *ejido*, and the authoritarian character of the planning process, Fox emphasizes the role of regional rural elites, known as *caciques*, in achieving "rural underdevelopment," which he defines as "the process by which a small minority dominates the key decisions which shape the lives of the rural majority via concentrations of social, political and economic power."[82]

Since *caciques* regularly operate at the regional level, "influencing the local implementation of national development programs including road building, credit allocation, and irrigation construction," Fox calls attention to the importance of *caciquismo* in shaping the underdevelopment of the region and the "region's integration into the national political system."[83]

Since the approval of the 1992 Article 27 was propelled by private interests both national and international, it is possible to assert that those national regional elites have contributed to shaping a new landscape in rural Mexico. The next chapter will concentrate on the role and meaning of those forces in transforming Article 27.

In spite of the pressure from national and international interests to decide the fate of rural Mexico, forms of organizations and resistance are still taking place in rural areas. The uprising in Chiapas contradicts Fox's claims of "minimal" opposition. On January 1st, 1994, a guerrilla group calling itself *Ejército Zapatista de Liberación Nacional* (EZLN) or National

Liberation *Zapatista* Army, took over several cities in Chiapas in protest to the enactment of NAFTA, and to declare war against the Mexican government. The *Zapatistas* of the 1990s denounced NAFTA and the restructuring adjustment policies that accompanied it, as the instruments of the Salinas - administration and supranational institutions working to destroy Indigenous cultures and their way of life. The three main demands of the EZLN were land, work, and housing, all of them related to land use and ownership. The movement initiated by the *Zapatista* Army questioned the "consensus" for the policies implemented to "modernize" Mexico and contradicted the assumed childhood that Herminio Blanco, principal negotiator of NAFTA, had condescendingly assigned to Mexican rural communities.[84]

CONCLUSIONS

Arguments for and against the New Article 27 depicted the contradictory meanings of land ownership, of demands obtained during the 1910 Revolution, of notions of nationalism, and of development trajectories that Mexico should follow. Those arguments mostly targeted the contradictions of Article 27 such as the innumerable exceptions to the law and the way it was used by the State to politically control rural populations. References to its origin, history, and evolution were rarely touched upon in those pro and con arguments.

The 1917 Article 27 was about land, land use, and its ownership as it benefited Indigenous communities and their descendants and, as general as it might appear, for Mexican-born populations. It reflected a nascent nationalism that wanted to reclaim land resources from various colonizing elites who had voraciously appropriated and unproductively accumulated Mexican land and natural resources.

The 1992 Article 27 is also about land and ownership, but both the beneficiaries and the potential land uses have changed. Legally, no more land redistribution will take place, no more dissolution of *haciendas*, no more restitution of lands violently appropriated, and no more protection of *ejido* and communal lands. Restrictions on national origin are gradually being lifted or turned around. The new Article 27 denies the history behind land use and rural communities in Mexico.

What was the justification for the approval of these changes? Basically, it was claimed that there was no more land to redistribute. However, *latifundios* still exist, indicating that there must certainly be some land available. During the 1990s, arguments about shrinking the role of the State and turning its activities over to the efficiency of the private sector predominated. Ironically, however, by implementing the 1992 Article 27, the State actually created new bureaucratic institutions. I will elaborate further on this in the next chapters. For now it is important to keep in mind the fact that the land-measuring process, which in more than 75 years of Land Reform the State was unable to execute, was accomplished in only a couple of years by these new institutions. The goals, of course, were differ-ent. The 1917 Article 27 required measurement data in order to return and redistribute land; the 1992 Article 27 required that data in order to parcel and privatize land.

Privatization of land belonging to Indian and other rural populations has been taking place since the era of the Spanish Crown. A profound racism toward Indian communities and an immeasurable greediness have been enduring characteristics of the large landholder elites in Mexico. Those elites seldom used land and natural resources to contribute to any economic development other than a feudal type of exploitation, or simple land accumulation. This behavior can, in fact, be seen as a defining charac-teristic of rural economic elites in colonized countries.

The statement by Tannenbaum cited on page 24 in this chapter is still relevant. Current policy makers still consider Indigenous communities a burden to the national economy and will continue to do so as long as the national economy is defined as an economy free of Indigenous communi-ties. The latter have historically been considered obstacles to whatever the modernization project is being promoted at the moment. The racist as-sumptions about Indigenous groups that shaped the colonial era continue to this day through the "exceptions" included in the 1917 Article 27 and later with the 1992 *Ejido* Reforms.

The mechanism to remove and exterminate Indigenous communities has been the privatization of their territory. Although the redistribution ele-ment of the 1917 Article 27 was not fully carried out, the legislation was very progressive and, at least theoretically, provided legal protection for Indigenous lands and those of other rural populations. The fact that this legislation is no longer in place threatens the very life of these peoples; it accelerates ethnic cleansing in Mexico. The threat that *Zapatistas* in Chiapas foresaw in the passage of NAFTA was that their way of life and, even they themselves, would eventually be destroyed.

Many Mexicos exist in a single territory. The regional differences out-lined in this chapter about the competing agendas clashing during the 1910 Revolution demonstrate this fact. The regions examined in this chapter were and are different in class, ethnicity, and geography. Both Tannenbaum and Katz's analyses call attention to the importance of regional history and culture, which the drafting and implementation of macropolicies constantly ignore. In 1951 Tannenbaum observed:

> These separate cultures vary greatly in their relative importance in differ-ent parts of Mexico. The northern part of Mexico is *mestizo* and therefore more European; the southern part is largely native and therefore more Indian in both race and culture.[85]

This is not to say that populations have not changed through time. Intra and inter migrations have changed the composition of the populations in the Northern and Southern regions. However, regional history and that of their original populations still play a role in their development trajectories. Because of that history, Katz acknowledges, the agenda of the *Criollo* and *Mestizo* populations of the North was radically different from that of the Indian and *Mestizo* populations of the South. Economist Daniel Coss puts it succinctly, "When one talks about *ejidatarios* in the North, one is referring to well-off landowners who drive a pickup, wear $500 dollar boots, with access to credit, with resources. When one talks about *ejidatarios* in the South, one is referring to poor *campesino* Indians who do not even have a mule, wear huaraches and have no access to credit."[86]

The 1917 Article 27 was a social contract that regulated a very unequal relationship between State and rural populations. Because it protected different forms of property, the most powerful advocates of private property constantly insisted on exceptions to the detriment of communal and *ejido* lands. In most cases, exceptions favored upper and middle landlord classes and capital, while disadvantaged Indigenous communities often suffered.

State repression in rural areas has been common in Mexico. Though traditionally portrayed exclusively as paternalistic, the Mexican State has in reality encountered strong resistance by rural communities who have opposed a ruling elite accustomed to seeking their votes, but ignoring their claims. At times, the State succeeded in co-opting some leaders of those rural communities. That co-optation indicated and reproduced internal hierarchies and power relations within the *ejido* that coupled with external constant repression.

Repression has followed displaced and landless communities as they migrated into urban areas and occupied the periphery of cities. As urban areas grow, their peripheries invade the nearby *ejidos*. Hence, those urban peripheries have been composed of what once were *ejido* lands. Today the rural migration into cities represents a transformation of the *ejido* and its communities, from peasants into urban workers. They continue fighting for land, not to work it, but to simply have a place to settle and live. Article 27, in its 1917 and 1992 forms, has had broader implications for land use, for access to affordable urban land, and for the development trajectories in Mexico's urban regions. The next chapter will deal with the urban and regional dimensions of Article 27 and the land use conversion of *ejidos*.

Chapter Three

Urban and Regional Dimensions of Article 27

Ejido lands have played a determinant role in the formation of urban areas and of economic regions in Mexico. Historically, *ejido* lands have provided, willingly or not, most of the space on which processes of urbanization and industrialization have taken place. Those processes have demanded the privatization of *ejido* lands and their subsequent conversion from agricultural into other uses, i.e., to house people—labor force—and to locate economic activities. The *ejido* and Article 27 have been deeply implicated, not just in rural agricultural land use, but in urban and regional land use development as well. The task of this chapter is to examine the urban and regional dimensions of Article 27 and the privatization of *ejido* lands that has occurred as a result of those dimensions.

It is also the task of this chapter to emphasize that although the *ejido* system—included in the 1917 Article 27—was created to carry out Agrarian Reform for the benefit of rural and Indigenous communities, it also benefited semi-urban and urban populations. This chapter explains how the conversion of *ejido* lands to urban use provided affordable land to low-income groups in urban areas. Most of those populations were landless Indians and poor *Mestizos* who had been displaced from their rural lands and migrated to cities, but once there could only afford to live on the urban periphery. Because of increasing urban growth, cites had gradually enveloped the surrounding *ejidos*. The urban periphery has been comprised of *ejido* lands.

In Chapter Two, I explained that the *ejido* system was supposedly created to benefit the landless and dispossessed Indigenous communities and rural populations. This chapter, however, shows how in reality the *ejido* system benefited the Mexican economic elites. Since the State had the right to expropriate *ejido* lands in the name of the public interest, in many cases such appropriation was manipulated to advance the private interest of those elites. After the 1910 Revolution and according to the *sexenio*[1] in turn, the public interest called for regional economic development. Such policies were implemented when the Mexican government expropriated *ejido* lands in order to promote industrial, tourist, and housing activities among other economic functions. This chapter discusses regional economic development policies which, hand in hand with policies of land redistribution, directed investment toward Mexico City and the Northern region, but not toward the South. My objective in laying out those policies is to expose how they promoted regional and urban transformations of land in Mexico, and subsequently of the *ejido*. Furthermore, through those policies, this chapter reveals the authoritarian character of the State when implementing land use planning at the urban and regional levels.

A further objective is to point out the economic forces and social actors intervening in *ejido* land use changes. In comparison to *ejidos*, private property was hardly affected because, over the years, defenders of private property had craftily included in Article 27 several protective mechanisms that made difficult its expropriation. The State turned instead to the *ejido*, and expropriated, privatized, and converted *ejidos* into other uses. The Mexican State was not the only actor intervening in the privatization of *ejidos*. National and foreign capital, land developers, *ejidatarios*, rural migrants, and urban populations in all income brackets intervened in the legal and illegal privatization of *ejidos*. Consequently, this chapter demonstrates that the *ejido* system had an impact beyond the rural arena and the agricultural sector.

Like the rural *ejido*, the "urbanizing" *ejido* has also been the scene of struggles between different ethnic groups, classes, cultures, and economic interests. Had it not been for the *ejido* system, the urban and regional geographies of Mexico would certainly have taken a different turn. In a way the *ejido* system worked as a growth control mechanism because it did not permit the legal sale of *ejido* land.

In Chapter Two, we saw that the transformation of Article 27 in 1992 restructured the relationship between the State and the countryside. Because the *ejido* has been affected by urban and regional land use development, it can be advanced that the 1992 Article 27 also restructured the relationship among the State, urban areas, and regions. The 1992 *Ejido* Reforms or 1992 Article 27 influenced the formation of urban areas and of economic regions in Mexico and these are the dimensions that this chapter explores.

DISPLACEMENT TO THE PERIPHERY

The history of colonial urbanization and industrialization in Mexico is related to the earliest dispossession and privatization of Indigenous and rural lands (see Chapter Two). It is important to emphasize, however, that "the urban" and "the city" did not originate with Spanish colonization.[2] In fact, when the Spanish arrived, vibrant Indigenous cities were already in place, with a strong organization of bureaucratic and planning activities.

In the case of Mexico City, these activities included the division of urban space by the Aztec State. The land of the Aztec city was planned into *barrios* or *calpullis* which comprised land for housing and for bureaucratic, religious, and commercial functions. In this way, the Aztecs assigned land for "a communal center with its palace or *tecpan*, a temple or *teocalli*, and a marketplace or *tianguis*."[3]

The Spanish conquest brought extensive destruction of most of the Indian cities in Mexico and disruption of their economic activities, both in the countryside and in urban areas[4] Most Indian populations were dispossessed and displaced from the land on which they had lived and from the built environment they had created. The case that epitomizes the effect of the Spanish conquest on the built environment of Mexican Indigenous cities is that of Tenochtitlán. This city formerly occupied the space that is now Mexico City. At the time of the Spanish conquest in 1521, Tenochtitlán had a population of at least 300,000 inhabitants.[5] After the conquest, the Spanish buried Tenochtitlán under an enormous plaza that is now the Zócalo, the main plaza in downtown Mexico City. The stone blocs of some of the pyramids of the Aztec city were used to build colonial architecture. Some of those sculptured stone blocs can be identified even today in some of the colonial buildings in downtown Mexico City. The process of destruction of the Indigenous cities[6] by the Spaniards and of building colonial cities on their ruins was only the beginning of the continuous destruction of Indigenous spaces and their substitution by colonial spaces.[7]

With the displacement of native Indigenous populations from their original cities, those populations moved to lands peripheral to recently-built colonial cities.[8] Here they settled and started working for the Spanish who were living within the city and who required Indigenous products and services.[9] This spatial arrangement defined the Spanish core and the Indian periphery within many colonial Mexican cities, an arrangement that in many cases subsists to the present day.[10]

THE EMERGING CITIES AND REGIONS

The Spanish conquest also restructured the economy of Indian cities and regions. Most of the economic activities initiated by the Spanish were of an extractive character.[11] Mining, commercial agriculture, and the textile industry were organized as part of the new transatlantic trade between America and Europe.[12] In a way, the relationship between the Indian periphery and the Spanish core of colonial cities was reproduced at the global level with the colonial territories representing the Indian periphery, and Europe representing the colonial core.[13]

In colonial Mexico, the extractive economic activities developed cities, most of them in the Northern region. Guanajuato, Taxco, Pachuca, Saltillo, Zacatecas, San Luis Potosí, and Durango became lively mining cities. As part of the Bajío region—located between Guanajuato and

Querétaro—Salamanca became a prosperous agricultural city. Veracruz in the Atlantic coast and Acapulco in the Pacific grew into key commercial city-ports. It is important to emphasize the role of these economic activities in their relation to the land that they required because their need for lands provoked continuous displacements of Indigenous populations from their remaining territories.[14]

Privatization, dispossession, and displacement continued after the 1810 war of Independence.[15] Although during this period mining started to decline, new lands were needed for the establishment of the textile industry and the subsequent expansion of commercial agriculture. These activities started shaping cities and regions in the areas where they developed. In the same way that Tenochtitlán was the primary city of its time, during the colonial and post independence periods, Mexico City became the primary city in Mexico. It concentrated not only administrative and political functions, but also provided access to different regional markets. Over time, the primacy of Mexico City would exacerbate and contribute to the concentration and centralization of activities, as well as determine the way other regions and cities grew.

THE REGION THAT CONTRIBUTED TO
DEFINE THE NORTH FROM THE SOUTH

The subsequent development of cities and regions did not vary significantly until the 1810 War of Independence. During this armed movement, the shaping of Northern and Southern regions as indicators of Spanish and Indigenous spaces, respectively, was further evident. The effects of those ethnic and economic segregations have somehow remained to this day, in spite of the massive migration of Southern populations to the North.

According to Eric Wolf in his study of the Bajío region in Mexico,[16] previous to the 1810 Independence movement, the North and the South developed different regional characteristics, due either to the lack of or the availability of an Indigenous labor force. The Indigenous labor force in the North had been almost exterminated; therefore, the Spanish in that region were forced to develop new ways of organizing their economy and, according to Wolf, they became more independent. Meanwhile, the Spanish in the South depended heavily on the Indigenous labor force, causing the South to develop a more dependent regional personality.

Wolf uses this comparison to find an explanation for the presumed entrepreneurial spirit of the Northern elites in comparison to those of the South. He also attempts to find an explanation for the different ways in which Northern and Southern elites related to their labor force. Wolf fails to indicate, however, that the Indigenous labor force was involved in economic activities related to land. Thus, the extermination of an Indigenous labor force liberated their land and resources and was, in fact, the element that contributed to making the Northern elite more "independent." The South had also exterminated Indigenous populations, but to a lesser degree. In order to survive in the city the urban elite needed Indigenous labor and Indigenous products, thus, the degree of extermination was not as high as in the North. Wolf also fails to point out that the Spaniards who settled in Mexico City were very different from those who traveled to the North. Their culture and their resources, even before their arrival to the Americas, were distinct.[17] In spite of those oversights, Wolf's conceptualization of the different cultures in the North and in the South is important if we are to understand the nature of these two different regions.

Between the North and the South lay the Bajío, which was like a "frontier" region. It was unique because of its economic diversification and its geographic proximity to Mexico City. The combination of mining, agriculture and the textile industry fostered favorable commercial links not only with Mexico City, but also with Guadalajara and Acapulco on the Pacific coast. Mining was the dominant industry and agriculture developed around it to support the labor force and the animals that worked in the mines. The Bajío region required a stable agriculture that could permanently support the mining sector. It was the motive behind the development of the area's irrigation system, a system that would become one of the causes of the region's prosperity.

According to Wolf, the scarce Indigenous labor force existing in the Bajío region managed to integrate itself into the regional economy and to achieve social mobility.[18] This mobility attracted different Indigenous communities from Central and Southern Mexico whose migration into the Bajío increased the population of the cities and the ethnic diversity of the region. Up to a certain point, Spaniards, *Criollos*, *Mestizos*, Indians, and Blacks shared the same space. This latter group had been brought to work in the mining sector as slave labor and later mixed mostly with Indian populations.

Around 1766–1767, the region witnessed the formation of group alliances among different ethnic groups who either rebelled against the working conditions in the mines and/or against the exorbitant taxation imposed on them by the Spanish colony. Their uprising was repressed and discriminatory labor laws were enacted to control Indigenous and Black communities. These laws required the resegregation of Indigenous and Black populations and the dispossession of their land. They even prohibited Indians and Blacks from dressing in Spanish-style clothing. The evident purpose of these laws was to deprive Indigenous and Black populations from the mobility they had achieved and to segregate them spatially, economically, and ethnically. Their brief "integration" and social mobility encountered fierce opposition as they tried to cross the racial, ethnic, economic, and social boundaries set by the Spanish Crown. These boundaries well defined the Northern from the Southern region.

The labor force in the agriculture sector was very diverse. Indians, Blacks, and poor *Mestizos* worked the *haciendas* of the Bajío Spanish elite. In contrast, the wool textile industry suffered shortages of labor since, by law, Indians were not permitted to work in this industry. Because of its smaller size, the industry could not afford Black slave labor. Ethnicity played a determinant role in the location and condition of the labor force as Indians and Blacks could not participate in all labor markets. In addition, they were obliged to pay taxes that Spaniards were not.

The labor force in the mining, agriculture, and textile industries in the Bajío was very diverse. This diversity gave rise to very independent social actors who, according to Wolf, brought about political alliances that later resulted in the Independence movement of 1810—a movement that originated precisely in the Bajío region. Both the economic elites—*Criollos* and well-off *Mestizos*—and the labor force—Indians and Blacks—wanted to change their relation with the Spanish Crown, although individually they held very different group interests and perspectives.

REGIONAL DIFFERENTIATION AND THE
URBANIZED INDIGENOUS LABOR FORCE

The importance of providing a historical perspective on a region such as the Bajío is to further understand the regional differentiation in Mexico between the North and the South. As explained in Chapter Two, this differentiation involves ethnic and class struggles that manifested in opposing political agendas regarding land tenure and its redistribution. This chapter insists that those struggles are evident also in regional development and urbanization policies.

In Chapter Two, I indicated the ways in which land tenure and land policies contributed to the containment or expulsion of the Indigenous and rural population from their lands. As land tenure changed, population remained or moved out. One result of those policies was the process of migration that sent rural populations into urban areas. This process expanded the labor force available in urban areas. In addition to proletarianizing the urban labor force, those land policies unintendedly contributed to the formation of cities and regions.

In this sense, I sustain that Mexican cities are "cities of peasants"—as Robert proposes to explain the urbanization of the Latin American city.[19] Because those who migrated from rural areas were mostly dispossessed Indigenous people and rural communities, the racism that they had faced in the countryside resumed in cities. Although the places of residence and the economic activities of those displaced immigrant populations had changed, the racism directed at them continued once in the city.

The history of the policies that shaped urban and regional spaces in Mexico seems to be characterized by a steady ethnic discrimination against Indigenous and other rural communities and by a policy bias for supporting Mexico City and the Northern region where most of the economic and political infrastructure began concentrating. These elements will be highlighted in the account of the regional and urban policies in Mexico that follows.

THE REGIONS AND CITIES UNDER DICTATORIAL PLANNING

During the dictatorial regime of Porfirio Díaz or *Porfiriato* (1876–1911), the economic base of regions continued to be organized to serve the needs of the world market as it had been during colonial times. New and more extensive and intensive displacement of rural populations took place as railroads, *hacendado* landlords, and new waves of mining activity took over land throughout Mexico.[20] According to Johns, during this period Mexico City was the "central place" that absorbed most of the expelled rural migration. It sheltered "two hundred thousand peasants who had lost their lands, but not their country ways."[21] Most of those peasants were of Indigenous origin.

Although Mexico City was the location that drew the majority of rural immigrants, other economically important regions such as Guadalajara, Monterrey, and Chihuahua also received displaced rural populations. Almost none of those populations was incorporated into the local industry, a few went to work for the commercial agricultural sector, some found under-subsistence petty jobs, and many had started settling in the undeveloped periphery of those urban centers. Historically, then, Mexican cities have been "cities of peasants"[22] or, to be more specific, cities of mostly dispossessed and displaced peasant Indigenous communities.

The process of restructuring of regions and cities in the *Porfiriato* times was not new nor was the constant violent "grab" of Indigenous lands and their subsequent concentration in a few private hands. As Johns states:

> The countryside was *the* major source of wealth for the city and its ruling class. By the late 1880s, a few thousand families had acquired one third of Mexico's land through a combination of lawful purchase, legal chicanery, and violence. Over half of all Mexicans lived and worked on *haciendas* whose owners often took their rents and profits to Mexico City.[23]

We can venture that the same process took place in other economically important cities, since Scott states that "by 1910 some 90 percent of Indian villages in the Central plateau had no common lands."[24]

Porfirio Díaz exercised a dictatorial planning that concentrated investment and infrastructure mostly in Mexico City, which reinforced it as "the core" city-region of the country. A few selective cities were strategically connected by the railroad system to the core-city. Railroad transportation contributed to the economic development of certain regions and cities as it permitted the transportation of export goods to European and American markets. However, the railroad also contributed to the destruction of Indigenous communities who were deprived from their lands, in order to build the railroads. The Porfirian government seized and privatized Indigenous lands to house the new economic spaces created under this dictatorship. Indigenous lands became the sites where mining, commercial agriculture, or industry settled. These industries increased the prosperity of the Porfirian economic elites, while for rural populations those industries meant the destruction of their local economies.[25]

Mining, agriculture, and manufacturing in Nuevo León, Chihuahua, Sonora, and Coahuila continued to craft the character of the Northern region and of the urban centers in which they were located.[26] The mining industry was the driving force that developed the Northern region. As this industry grew, other economic activities, such as manufacturing and agriculture, developed to supply the needs of the mining industry and of the increased population in northern cities.[27] As railroads and commercial agriculture needed more land, the prosperous agricultural Bajío region, one of the agricultural suppliers of the Northern area, displaced more peasants in order to extend its agricultural fields. New waves of immigrant peasants descended upon Mexico City to become the new urban labor force. "Women worked as domestic workers, nannies, and seamstresses; while men began working as ambulant merchants, artisans, day laborers, street cleaners, brick layers and peddlers."[28]

Social class and ethnic background defined the urban strata. The aristocracy was white and Spanish, either from Spain or born in Mexico of Spanish descent. The middle class was mostly *Mestizo*, and the working classes were composed of poor *Mestizos* and Indians. Johns cites in regard to this latter group, the racially charged work of Julio Guerrero who described several ways of classifying *pelados* or working and underemployed populations in Mexico City. The first group was comprised of *Mestizos* and Indians who had migrated to the city, learned Spanish, and adopted city working class clothing styles. According to Guerrero, this first group was immoral and corrupt. The second group was a wave of migration that had kept their Indian traditions and lived in Indian neighborhoods, or on the periphery of the city. Finally, the last group was also comprised of *Mestizos* and poor Indians similar to the ones described in the first group. However, they lived in *vecindades*[29] and belonged to the urban working class. No matter how they were classified, they were still looked upon as *pelados*, "the dirty and nauseated people" as opposed to "the polite classes" that were the white and well-off Mexican elites.[30] Despite the passage of time, some of these same characterizations still remain in place in contemporary Mexico.

The combination of inherited racism, dictatorial planning, and transfer of capital and labor from export-producing regions caused certain cities to develop more than others, especially when foreign investment in industry was initiated in urban areas. This was the period in which public utilities flourished to improve well-off neighborhoods. But the same flourishing did not occur in those areas, within the same city, where expelled peasants had relocated. Scott describes: "The government ignored several petitions for paving, water, and sewers, because the settlement, which emerged in the 1900s, was not authorized."[31] As a historical curse, this illegality and lack of services that spun off from the racism that plagued planning would become persistent characteristics of the squatter settlements in Mexican cities. Those were the urban and regional circumstances that the Agrarian 1910 Revolution encountered and that persisted during and after Agrarian Reform. Indeed, as war further displaced Indigenous and rural populations, they were forced to migrate into cities, contributing to the growth of the urban periphery.

LAND REFORM, RURAL MIGRATION, AND STATE INTERVENTION AFTER THE 1910 REVOLUTION

The end of the 1910 Revolution brought the enactment of Agrarian Reform through Article 27 and the organization of *ejidos*. Agrarian Reform promised to eliminate those land monopolies called *latifundios* and to redistribute their lands in the form of *ejidos* to the landless and dispossessed. Since land is both an economic and a physical resource, Agrarian Reform would potentially initiate a new stage in the rural, urban, and regional processes in Mexico. A new redistribution of resources would take place along with a new configuration of regional and urban spaces.

The first administration in charge of applying Article 27 was that of Venustiano Carranza (1915-1920). As mentioned in Chapter Two, Carranza represented the most conservative agenda in the 1910 Revolution; he had opposed Agrarian Reform and advocated the protection of private property. As a result, during his tenure no major land redistribution took place. After his term, five different presidents took over during the period 1921–1934. All of them belonged to the so-called Sonora group[32] and, like Carranza, they did not affect significantly the *hacienda* lands of the Northern elite from which the Sonora group descended. From 1917 to 1934, a seventeen-year period, 110,234 square kilometers were distributed to 947,526 peasants[33] (see Table 3.1).

TABLE 3.1 Land Redistributed by Pre-Cárdenas Administrations and by Cárdenas, 1917–1992

Administration	Period	Land Redistributed (sq. km)	Total Beneficiaries
Pre-Cárdenas			
Carranza	1917–1920	3,819	77,203
Obregón	1921–1924	17,307	164,128
Calles	1925–1928	31,863	302,539
Portes Gil	1929–1930	24,385	187,269
Ortiz Rubio	1931–1932	12,258	57,994
Rodríguez	1933–1934	20,602	158,393
Subtotal	1917–1934	110,234	947,526
Cárdenas	1934–1940	201,459	764,888

Source: Zaragoza, J.L. and R. Macías, *El Desarrollo Agrario de México y su Marco Jurídico* (México, D.F. Centro Nacional de Investigaciones Agrarios, 1980). Based on Table 8-1 of that book.

The slowness of land redistribution increased the rural migration into cities and the subsequent pressure on urban housing. Land speculators and real estate developers appeared in the urban land market. As the demand for housing increased, so did the abuses of speculators and developers. This situation brought forth protests from the working class and even from different sectors of low and middle class bureaucrats. Those protests drove the State to intervene in urban and housing policies, and consequently in the urban land markets.

The post revolutionary period marked the emergence of different political and social arrangements between the Mexican State and the swelling working class, most of them urbanized Indians. The State intervened in housing policy by expropriating rural and Indigenous lands in order to plan and build several working class neighborhoods or *colonias proletarias*.[34] These planning and investment actions brought the State into the urban land market and established a long-term political relationship between labor unions and the State. The main beneficiaries of *colonias proletarias* were postal, railroad, textile, and blue collar workers as well as bureaucrats.[35] Most immigrants from the countryside living in the periphery of Mexican cities did not benefit from this State intervention.

Three other factors contributed to rural migration and regional change during this period. The first factor was the development of the highway system in the 1920s. The system was built with the purpose of connecting major regional centers and cities among themselves and with Mexico City, but their construction once again displaced people from their lands.[36] Another restructuring factor was the expansion of the manufacturing sector, stimulated by the investment of national capital. Because of the revolution, foreign capital fled, and the remaining capital was reluctant to contribute to the post-revolutionary economy. The oil industry, for example, continued operating and growing in Mexico during and after the revolution, however, its profits were repatriated to either the United States or to England thereby lessening their economic contribution to the Mexican economy. A final element contributing to rural migration in this period was the 1930 world economic crisis that negatively affected both mining and commercial agriculture.

The post-revolutionary period witnessed a new economic restructuring of Mexican cities and regions. Because of the world crisis, mining cities and commercial ports lost momentum. Meanwhile, the same highway system that allowed "the location of new manufacturing activities in cities with large market potentials,"[37] served to increase the movement of rural populations into the cities. The factors that contributed to the exacerbation of rural migration into cities were in fact the same factors that further defined the economic superiority of the Northern region, i.e., highways, manufacturing, and investment.

AN ATTEMPT FOR PROGRESSIVE PLANNING?

During the Lázaro Cárdenas administration (1934–1940), Agrarian Reform was finally carried out effectively. The total amount of lands redistributed surpassed those previously allocated by the presidents of the Sonora group. Cárdenas[38] distributed in six years approximately 82 percent more of what Carranza and the Sonora group had distributed in the seventeen years of those administrations (see Table 3.1).

As *haciendas* were divided and redistributed in the form of new *ejidos*, the Cárdenas administration supplied needed credit and technology to support the countryside. The *Banco Ejidal* or *Ejidal* Bank was created and irrigation projects, roads, and ports were built or revitalized.[39] *Ejidal* agriculture became a strong axis of the Cárdenas administration. All over Mexico *ejidos* were created, but their extension and the economic and technical support that they received varied according to the region. The *ejidos* in the North were larger and received more support than those in the South.[40] Nonetheless, the impressive implementation of land reform during the Cárdenas tenure decreased the rural migration into urban areas[41] and the rate of urbanization of cities.[42] The Cárdenas administration was an attempt for progressive planning.

Land redistribution and the expropriation of the oil industry in conjunction with the world-wide Depression negatively affected the economy of some of the Northern regions, but increased the well being of those populations which had benefited from the return of their lands. Afraid of the progressive character of the Cárdenas administration, national and international capital took revenge by limiting investment and credit, by transplanting national capital to regions outside Mexico, by threatening intervention from the U.S., and by colluding with fascists and rightist groups in Mexico.[43] These factors had a regressive impact on the Mexican economy and as a result the wages of the urban population suffered.

The economic model that Cárdenas wanted to advance was based on the *ejido*, with industry being subordinated to it.[44] Paradoxically, the construction of an infrastructure and the significant investments in human capital that he promoted, also helped national industrial capital and laid the base on which import-substitution industrialization would take place. In this regard, Brachet-Marquez points out that these investments, in conjunction with the nationalization of the oil industry, greatly benefited "private banking, industrial production, and commercial agriculture [. . .] as evidenced by the 25 percent increase in industrial production between 1934 and 1938."[45]

Cardenismo was the period in which the incipient building of urban political institutions, which had started during the administrations of the Sonora group, developed even further. Workers in rural and urban areas were organized into major political associations. In time these political entities would become an integral part of the political apparatus of the official party, the *Partido Revolucionario Institucional* or Institutional Revolutionary Party (PRI), which would mobilize those groups for its political convenience in exchange for favors in kind, such as informal land or political power. The contradictions of the Cárdenas administration were

TABLE 3.2   Urban and Rural Population in Mexico, 1900–1990

Year	Urban	Rural
1990	28.3	71.7
1910	28.7	71.3
1921	31.2	68.8
1930	33.5	66.5
1940	35	65
1950	42.6	57.4
1960	50.7	49.3
1970	58.7	41.3
1980	66.3	33.7
1990	71.3	28.7

Source: Estadística Históricas de México, 1985: INEGI, 1992a. http://www.inegi.gob.mx

complex. He tried to implement a more progressive agenda and built progressive organizing institutions, but later administrations would misuse that agenda as a false promise to meet the demands for land and justice of both rural and urban populations. In the same vein, the PRI- State would manipulate the organizing institutions created during the Cárdenas period to set a clientelistic pattern between the State and the leaders of those organizations. This clientelistic relationship strengthens during the industrialization process of the 1940s and the accelerated urbanization processes that accompanied that industrialization.

INDUSTRIALIZATION AND URBANIZATION DURING WORLD WAR II

In the 1940s, Mexico's urbanization accelerated as a result of exterior conditions and interior regional economic policies (see Table 3.2). The outbreak of World War II was an important exterior condition that influenced that process of urbanization because it encouraged the return to commercial export agriculture needed to supply food for the U.S. Simultaneously, this economic restructuring opened the opportunity to finance, at the national level, the import substitution industrialization process that the Ávila Camacho administration (1940–1946) was promoting.[46]

World War II brought a change in the investment patterns and in Mexico's economic development policies, affecting both national and local economies. Steel, iron, copper, and coal industries were promoted as State enterprises and received extensive public investment. As mentioned earlier, the mining sector was located mainly in the North, thus, the aid provided to that sector gave further support to a region that historically had always been privileged.

The war represented the exterior element that influenced the domestic allocation of economic resources. As a result, land redistribution and investment patterns for agriculture were modified. While investment in *ejidos* continued at a constant of 735 pesos, investment in private commercial agriculture increased to approximately 1,164 pesos. The regional economic policy was evident; the support was going to export-oriented agriculture, mostly situated on privately-owned land in the North, versus subsistence or domestic agriculture, located mainly in the South. The support of private commercial agriculture decreased land redistribution because the State was not going to affect the land interests of the private producers. As shown on Table 3.3 the amount of *ejido* land allocated during this period decreased dramatically from "20 million hectares [201 thousand square kilometers] distributed under Cárdenas to nearly 6 million hectares [approximately 60 thousand square kilometers] under Ávila Camacho."[47] That accounted for about 70 percent less land redistributed. This economic panorama explains a new wave of growing urban population in Northern regions, and a new migration wave from rural into urban areas which negatively affected the rate of population growth in Southern regions.

These international, national, and local contexts explain the decline of the participation of the *ejido* in total agriculture production, "from 50 percent in 1940 to 37 percent by the end of the decade."[48] A renewed emphasis on export commercial agriculture is reflected in the enactment of the 1942 Agrarian Code. This law exempted from expropriation those private properties engaged in agriculture oriented toward the world market, providing those properties were smaller than 300 hectares. This amendment set a precedent that would bring about the *amparo* mechanism, approved in 1947, during the Alemán administration.[49] As mentioned in Chapter Two, these constant amendments were "exceptions" to avoid land redistribution and to protect private property.

TABLE 3.3   Land Redistributed After the Cárdenas Administration, 1934–1988

Administration	Period	Land Redistributed (sq. km)
Cárdenas	1934-1940	201,459
Ávila Camacho	1940-1946	59,704
Alemán	1946-1952	54,395
Ruiz Cortínez	1952-1958	57,717
López Mateos	1958-1964	93,082
Díaz Ordaz	1964-1970	230,556
Echeverría	1970-1976	120,171
López Portillo	1976-1982	56,089
De la Madrid	1982-1988	50,812

Source: Zaragoza, J.L. and R. Macías, *El Desarrollo Agrario de México y su Marco Jurídico* (México, D.F. Centro Nacional de Investigouones Agrarios, 1980). Based on Table 8-1 of that book and on data from the Secretaría de la Reforma Agratia provided by INEGI in Internet, http://www.inegi.gob.mx

During the 1940s, then, investment policies meant not only the support of certain Northern sectors, but also the privileging of those regions and certain types of land tenures. The combination of national policy and world circumstances continued strongly benefiting the mechanized agriculture of the North and Northwest, but not so the subsistence units of the South. Disinvestment in the South led to continued migration to cities, causing the urbanization of *ejido* and communal lands located near cities.

INFORMAL URBAN SETTLEMENTS IN *EJIDO* LANDS

Although the 1942 Agrarian Code bestowed *ejidatarios* with titles and certificates, those documents took years to obtain and in most cases *ejidatarios* just gave up on them. Cymet proposes an optimistic view of this amendment in the sense that the legislation specified "that the *ejido* is the property of the peasant communities, not of the state."[50] It was, however, from 1943 on that the State expropriated and incorporated large extensions of *ejido* land for urban housing as well as for manufacturing and other purposes. The pressure for land for urban housing continued to grow as migration from rural areas increased. The new immigrants could not afford housing in the city so they began to buy, rent, borrow, or simply invade *ejido* lands in order to temporarily secure a place to live. Since legislation did not allow the transfer of *ejido* land to non *ejidatarios*, these transactions were informally arranged or lands were illegally taken. Although such land parcels were illegal and lacked such "urban services" as drinking water, sewer systems, sidewalks, paved roads, and electricity among others, they provided a provisional living space for new peasant migrants who had lost their own lands and/or their agricultural jobs. This urbanization process took place not only in *ejido* lands near Mexico City, but also in *ejido* lands near other major Mexican cities, such as Monterrey and Guadalajara.[51]

EJIDO LAND USE CHANGES: EXPROPRIATION, EXCHANGE, AND *EJIDO* URBANIZATION ZONES

As noted in Chapter Two, the exceptions to Article 27 were multiple. One of them was the 1942 Agrarian Code. This code included three ways of allowing *ejido* land use changes: (1) expropriation; (2) *permuta* or exchange; and (3) the formation of *ejidal* urbanization zones. Although Ávila Camacho legislated land expropriations so that they could be affected exclusively for the benefit of the urban "popular" sectors,[52] expropriations for purposes other than the popular one, began during his administration. In this way, the agricultural use of expropriated *ejido* land was rampantly changed to industrial uses with the construction of the Vallejo Industrial Zone and the Azcapotzalco PEMEX refinery, which were located in expropriated *ejidos* adjacent to Mexico City.[53] These developments signaled the point at which future expropriations of *ejido* land would be justified to house all kinds of industrial and manufacturing sites.

The *permuta* or exchange mechanism consisted of bartering *ejido* land for private property or for *ejido* lands in other areas. For the most part, the lands exchanged were of lesser quality and disadvantageously located. The history of the exchange of *ejido* lands to build the exclusive area of the Pedregal de San Ángel in the South of Mexico City is a case in point. Here the State approved the acquisition of the Tlalpan *ejido* by calling on the exchange mechanism in the Code. As a result, Tlalpan *ejidatarios* received *ejido* lands in another area and were relocated to the State of Guanajuato. . . where they received land of poor quality and location.[54] Meanwhile, the developers of the Pedregal de San Angel acquired *ejido* land at minimal prices and the high-income residents who later occupied the houses built in this area had profited from acquiring cheap land in a well-located area of Mexico City.

Finally, the formation of *ejidal* urbanization zones was a mechanism already anticipated in the 1917 Agrarian Law. In a way, such zones provided a planning mechanism that anticipated future needs for housing for *ejidatarios* and for newcomers. In the law, these newcomers were called *avecindados*.[55] The legislation specified that *avecindados* could acquire *ejido* land as long as the *ejido* considered them "resourceful for their community."[56] Although the *ejidal* urbanization zones could not legally be used to expand the territory of neighboring cities, the fact that they accepted *avecindados* was an informal way of expanding the territory of cities.

Unfortunately, over time, the expropriation and exchange mechanisms were widely used, resulting in the incorporation of large extensions of *ejido* land into the private land market. The *ejidal* urbanization zones meanwhile were neglected, denying the possibility for *ejidos* to plan for and by themselves when to sell and whom to incorporate in their community as *avecindados*. Neglect of *ejidal* urbanization zones appears to be related to the speculative pressure that different actors had begun to exert on *ejido* lands.

In spite of all the pressures from the State, land speculators, political leaders, and people from different income levels, the *ejido* system worked in the urban sphere as an unintended growth control mechanism that kept urbanization and industrialization within certain limits. The *ejido* system slowed the conversion of agricultural *ejido* land into other uses.

INDUSTRIALIZATION AND INVESTMENT IN PRIVATE LANDS

Ávila Camacho's policies to promote national and foreign investment in the Mexican economy were continued during the Miguel Alemán administration (1946–1952). They led to further industrialization and investment, though not for the benefit of *ejidos*, but for the benefit of private proprietors. Alemán's economic policy strove to attract the capital that had left Mexico at the end of World War II when Mexico's commercial agriculture was no longer needed for the war economy. Alemán put in place further fiscal incentives and low wages, and "[b]y 1950, the capital that had fled between 1945 and 1948 was back, taking its cut from the devaluation and participating in a new industrial boom spurred by the Korean War."[57] The rapid industrialization was boastfully called the "Mexican Miracle" due to the impressive economic growth during those years. However, that economic growth was misleading as the countryside had been forgotten.

Agrarian redistribution during the Alemán period continued to decrease (see Table 3.3, presented previously in this chapter), while public investment increased and was a key variable for commercial agriculture and for the creation of State industries. These two latter sectors were highly successful due to the injection of public capital and the development of regional river basins in Central Mexico which, along with the investment in infrastructure, "absorbed 22 percent of the federal budget. But in this instance, the developed lands were not mainly *ejidales*, but private, and this was justified in the name of efficiency."[58]

The Mexican Miracle strategy drained financial resources not only from rural populations, but also from the urban working class, in the form of low wages and lack of social benefits. Those resources were transferred to private investors who benefited by paying low taxes and having access to credit from the Mexican government. In other words, the State financed industrialization and down-played social welfare.[59] This caused additional rural migration to cities at the same time worsening the basic living conditions for the old and new urban working classes.

It is evident that Mexico restructured its economy during the "Mexican Miracle" period, according to external conditions posed by the world economy and politics. As those external conditions changed, Mexico's internal relations and policies accommodated accordingly. In addition, the Alemán administration followed a strong industrialization policy that favored industry over subsistence agriculture, big landowners over *ejidatarios*, and the city over the countryside.

THE PROCESS OF INDEBTEDNESS

By the 1950s, the roller coaster development of the Mexican economy was already closely linked to the world economy, or at least to the U.S. economy. The end of the Korean War saw a new economic restructuring and a new wave of capital flight. This period initiated a process of indebtedness that continues to the present day. At that time, the administration of Ruiz Cortínez (1952–1958) followed in the footsteps of his predecessor Alemán. A new devaluation of the peso, new fiscal and credit incentives for national and foreign capital, and an additional freeze in wages were aimed at attracting national and foreign investment. This economic strategy was called *desarrollo estabilizador* or "stabilizing development." Through this policy, the State invested in those sectors in which private capital was scarce or absent. Those sectors became the well-known *paraestatales* or

State-owned industries, which directly or indirectly financed private capital by selling State-produced goods and services at subsidized prices.[60]

Different policies were implemented during the Adolfo López Mateos (1958–1964) period. During his administration land redistribution revived, wages increased, and social investment expanded. Next to Cárdenas, López Mateos was the president who distributed the largest amount of *ejido* lands to landless peasant communities (see Table 3.3, presented previously in this chapter). However, due to the economic and fiscal situation in Mexico at that time, the well-meant policies caused the national and foreign debt that financed them to climb. An expansion of welfare policies coupled with the "stabilizing development" strategy which continued offering fiscal and public incentives for capital investment, expanded hugely Mexico's foreign debt. A debt was easily contracted since the Alliance for Progress was offering easy loans to developing countries to promote industrialization. It was during López Mateos' presidential term also that the oil industry consolidated in the states of Veracruz and Tabasco, defining them as important oil producing regions.[61]

The final administration to implement this "stabilizing development" strategy was that of Díaz Ordaz (1964–1970). Under this administration, the support to private capital investment continued, but at the cost of ignoring social welfare. Foreign indebtedness also grew, except that the Alliance for Progress was no longer in place. In spite of restrictive social welfare policies, this administration made housing mandatory for workers. In addition, land redistribution was extensively carried out. It has been pointed out, however, that the housing policy was a limited preventive measure designed to contain growing labor demands, while land redistribution was promoted in order to placate the emerging guerrilla movements in rural areas such as Guerrero in Southern Mexico.[62]

GROWTH POLES IN *EJIDOS*

It was also during this era that the idea of "growth poles" became part of the regional investment strategy of the Mexican State. According to the "growth pole" theory, "poles" could be either firms or industries, or groups of them, which once located in an optimal geographic space could induce regional economic development.[63] The growth pole idea led to investment in such mega projects as the petrochemical industries in Southern Veracruz and tourist resorts in Acapulco, Guerrero and Veracruz in Southern Mexico.[64] Many of those poles were built on expropriated *ejido* lands.

This period also witnessed the first attempts to undertake regional planning through the development of industrial cities and parks. Ciudad Sahagún, built in 1953, was the first industrial city of its kind. Although Ciudad Sahagún developed with relatively little success,[65] its construction, along with other similar projects developed later, was an example of State intervention as a generator of regional economic spaces, as well as inducer of urbanization around those growth poles. These spaces were created mostly on *ejido* lands.

During the Díaz Ordaz administration, several industrial cities and parks were built with both public and private investments. Most of the private projects were located in Northern Mexico or around the Metropolitan area of Mexico City which reinforced the growing importance of the Northern region and the primacy of Mexico City.

THE PRIMACY OF MEXICO CITY

Although the growth poles strategy was designed to help develop other regions in Mexico and to address uneven development, it is evident that during the period 1940–1970, Mexico City had consolidated its role as the primary city-region in Mexico. It had gradually concentrated, in and around it, industries, services, and decision-making power. Territorially, it expanded over nearby *ejidos*, changing their uses either through urbanization or industrialization. Mexico City illustrates perfectly the unevenness of regional and urban development in Mexico. Because of its concentration and centralization, the capital attracted a considerable pool of labor as well as industry. As a primary city, it received more investment and public support than other city-regions, with the exception of tourist and oil-producing regions, and the North.

We can ascertain, therefore, the model of economic development followed by Mexico, by observing the centralized urbanization and the concentrated industrialization of Mexico City. As rural migration into Mexico City continued to increase the pressure on urban space, so land invasions of the nearby *ejidos* increased as well. *Ejido* land conversion in and around Mexico City laid the groundwork by which urbanization would take place in other major Mexican cities. There were, however, further attempts to address uneven regional development through decentralization policies.

DECENTRALIZATION AND NORTHERN ELITES

The economic development policies carried out during the Luis Echeverría presidential term (1970-1976) were an attempt to decentralize economic activities and to, supposedly, achieve a more equitable development. His *desarrollo compartido* or "shared development" program would increase both economic growth and social welfare. Those policies included fiscal reform, the increase of prices in the goods and services produced by State enterprises, and decentralization measures to promote social spending and industrialization in disadvantaged regions. This strategy, however, encountered internal and external pressures. Business regional elites like the Alfa and Monterrey Groups, both from Northern Mexico, strongly opposed the policies of "shared development." The attempts to address the uneven development that had occurred as a result of policies of previous periods were held back by those powerful regional vested interests that had been created by those very policies.

The decentralization that the Echeverría administration promoted implied a more intensive State participation in the economy. This participation called for the continuation of mega growth poles as a development strategy to diminish "regional inequalities."[66] "Industrial plants, tourist resorts, airports, harbors, and diverse urban infrastructure"[67] were built by the State in several Mexican regions. The State continued assuming the role of active participant in the shaping of regions and cities through the allocation of public and private investments in selective spaces. In this way, the Cancún tourist resort and the port and steel plant Lázaro Cárdenas—"Las Truchas"—were built in Quintana Roo and in Michoacán, respectively.[68] It is seldom mentioned that those growth poles were built on spaces provided by expropriated *ejido* lands. Nor is it mentioned that those expropriations, although carried out in the name of the public interest, displaced and dispossessed *ejido* and other rural communities from their living and working spaces.

By the 1970s, the gradual disinvestment in agriculture had taken its toll: agriculture was seriously declining. A chain effect occurred as agricultural imports increased to cover the internal supply that the local agriculture could not provide anymore. In turn, the price of those imports negatively affected the process of industrialization. In order to lessen the negative effects of the crisis in the countryside, Echeverría offered beneficial credit and insurance policies to small farmers. In addition, he expropriated approximately 100,000 hectares of land for redistribution. These were irrigated lands mainly located in the Northwest part of Mexico. These expropriations intensified the confrontation between the Echeverría administration and the private capital in that region.

According to Erfani, State policies during this period were constrained by constant "threats of capital flight and investment strikes" by private capital.[69] Those threats had a deep negative impact on Echeverría's agenda of "shared development" and decentralization. Thus, in 1976, when he proposed to legislate the real estate sector as part of his urban policy, he encountered bitter opposition. His urban policy would severely affect two of the most profitable investments of private capital in Mexico: speculative real estate development and tourism.[70] After several clashes, the State ended up approving a very tame law that did not affect the interests of big urban landlords.

EXPROPRIATING INSTITUTIONS AND PLANNING LAWS

Although urban policy during the "shared development" period was defeated, some important urban policy and planning institutions were created. Among them, the *Instituto Nacional para el Desarrollo de la Comunidad Rural y la Vivienda Popular* or National Institute for the Development of Rural Communities and Popular Housing (INDECO) and the *Comité para la Regulación de la Tenencia de la Tierra* or Committee for the Regulation of Land Tenure (CORETT) had a significant impact on the provision of urban low-income housing spaces through the regularization of *ejido* land that had been informally settled. These institutions were successfully created because their function was to "regularize" the *ejido* land that had for long been occupied informally.[71]

The INDECO was created in 1971, its jurisdiction comprised rural and urban spaces and its main purposes were to regularize land tenure, and acquire and plan *ejidos* for future urbanization. During the period 1971–1974, INDECO expropriated 7,900 hectares of *ejido* land near urban areas to be either regularized or retained as land reserve for future urbanization.[72] INDECO was restructured in 1977 when the CORETT took over INDECO's regulatory functions.

The CORETT was created in 1973 and since its creation it has been a key institution in the regulation and expropriation of *ejido* land where squatter settlements were informally established. In 1974, CORETT was upgraded from Committee to Commission and since then it has absorbed many of INDECO's former functions.

In addition to CORETT, the *Secretaría de Asentamientos Humanos y Obras Públicas* or Ministry of Human Settlements and Public Works (SAHOP) was created. This Ministry had the responsibility of applying the General Law of Human Settlements designed near the end of Echeverría's term in 1976. This Law required states to regulate, prepare, and administer urban planning. As part of this legislation several federal institutions were created to deal with the growing *ejido* invasions and the illegal nature of the new urbanizations mushrooming in the areas adjoining urban centers.[73] The creation of institutions such as INDECO and CORETT gradually expanded the intervention of the State in the planning and regulation of urban land, and in the transformation of the *ejido*.

THE URBANIZING *EJIDO:* CONVERTING *EJIDO* LAND TO HOUSING USES

The "urbanization of the *ejido*" established a new relation between the State, the *ejidatarios*, and the new immigrants settling in those *ejidos* close to cities. When invasions of *ejido* land took place, *ejido* populations demanded that the State quickly initiate expropriation, so that their land could be legally converted into urban land and they could be at least minimally indemnified. Since before 1992, by law, *ejido* lands could not be sold or transferred, the State had to follow a lengthy procedure to legalize or "regularize" those lands for conversion into urban lands. As a first step, the State expropriated *ejido* lands, then, it paid the *ejidatarios* for that expropriation. Once the State had become the owner of those expropriated lands, it could sell them to the squatters. Public officials called this process the "invasion-expropriation-regularization cycle."[74]

It was a different situation when *ejido* land was not invaded, but illegally sold. In this case, settlers were the ones urging land regularization which was to their advantage as the status of land as irregular prevented the introduction of urban services as well as access to housing credit. The process was similar to that already described; land was first expropriated by CORETT, then *ejidatarios* were paid for the expropriated land, and, finally the expropriated land was sold to and titled in favor of those who had originally bought it from *ejidatarios*. CORETT was a key institution in the regularization process. Although this process had not been defined by planning bureaucrats, it is evident that an illegal sale-expropriation-regularization cyclical process was taking place.

Both processes, the "invasion-expropriation-regularization cycle" and the "illegal sale-expropriation-regularization cycle," had pros and cons. In both processes, the State paid for *ejido* land based on its agricultural value, which was extremely low. However, it sold the land at its commercial value which prevented many squatters from having formal access to that land. This situation opened corners for land speculators to intervene and obtain a slice of the pie of the urbanizing *ejido*.

In the second process, payment was somehow compensated as *ejido* land was sold and bought twice. That is, *ejidatarios* first sold land to the new tenants. Then, when regularization took place and land was expropriated, the tenants had to repurchase the land from the State. In other words, the *ejidatario* sold the land twice, first to the tenant on an illegal basis, and then to the State as part of the expropriation. The tenant bought the land first from the *ejidatario* in an informal arrangement, and then legally from the State. Unfortunately, however, this also prevented some tenants from continuing to have access to the land they had bought from the *ejidatario*, due to their inability to afford the high prices set by the State.

The fact that those informal processes had allowed access to affordable *ejido* land for housing low income populations has been an argument for debate in the literature of the urbanizing *ejido*. The pro- argument claims that although invasion or informal sale were not legal means to obtain lands, at least they provided a way for low income populations to have access to housing that was not available to them by other means.[75] Those disagreeing with this argument state that the illegal sale of *ejido* land impoverishes *ejidatarios* since they do not receive appropriate compensation for land that legally belongs to them.[76] In reality, both arguments are true. *Ejidos* did provide affordable land for housing and most *ejidatarios* received minimum compensation for their selling their lands. At least, in the case of illegal sale, *ejidatarios* met their buyers, which did not happen in the case of invasions of *ejido* lands. In addition, some of those invasions were manipulated to secure political control.

THE PRI-STATE AND *EJIDO* LAND INVASIONS: A MECHANISM FOR POLITICAL CONTROL

Another mechanism to convert *ejido* land to housing uses is documented by Castells. He describes the role of PRI political leaders in organizing land invasions with the purpose of gaining some political leverage with the squatters who would later become their constituents. For Castells, land invasions in Mexico were a twofold process. On the one hand, invasions launched *ejido* lands into the urban market, on the other, they were used to socially control "people in search of shelter."[77]

In the case that Castells presents, it seems that despite PRI efforts, squatters refused to be "regularized." For one thing, the price of regularized land was prohibitively expensive to those settlers. In addition, some of them feared that regularization would bring gentrification and they were right. Once legalized and assessed at their commercial values, *ejido* lands stopped being affordable for those settlers who had self-constructed their homes, provided basic urban infrastructure, and participated in social movements to petition urban services from the State. Apparently, this was a cyclical process as some of the settlers, who could not afford regularized land, had to move to other lands to "take care of their housing needs"[78] and begin anew constructing their own city. Meanwhile, those who could afford the regularized land took advantage of the urbanizing work carried out by those initial settlers. From this perspective, it is possible to affirm that regularization was the first step in a series of gentrification waves that the urbanized *ejido* would undergo. Regardless of the life time and efforts that those settlers had invested on constructing their city, which made Castells characterize them as "the driving force in the social production of urban space,"[79] they faced the consequences of gentrification in the midst of the economic crisis of the late 1970s.

LA CRISIS: THE ROAD TO NEOLIBERALIST POLICIES

The economic crisis during the López Portillo administration (1976–1982) was partially created by the cumulative effects of the economic policies undertaken in former *sexenios*. As previously stated, those policies heavily subsidized private capital at the cost of labor. The crisis was also the result of the role of private capital that used its resources to speculate instead of investing in productive activities. Finally, the increasing foreign debt to cover public spending also contributed to the crisis.

The economy in this period briefly expanded due to the incorporation of new oil reserves. This oil boom petrolized the economy and developed new regions. Public investment went to oil industrial ports such as Salina Cruz in Oaxaca, Coatzacoalcos in Veracruz, and Tampico-Altamira in Tamaulipas. The first two ports are located in the Isthmus oil producing region in the South, while the last one is located in the North oil producing region in the Gulf of Mexico. But the oil boom was brief and the decline in oil prices at the end of the López Portillo administration provoked an even deeper crisis. Technocrats in the Mexican system conceived of the crisis as the result of State intervention in the regional and urban development of Mexico. They ignored, however, the external and internal factors and actors, other than the State, that also contributed to this crisis.

The crisis marked the beginning of the change towards neoliberalism and, consequently, towards a different direction in regional and urban policies. Neoliberalism advocated a shrinking role of the State in the economy, ample privatization of State-owned industries, and the reduction of the welfare State.[80] Basically, neoliberalism proposed a different allocation and organization of regional, urban, and rural spaces, which would affect Mexico's internal and external economies and relations.

It was also during the crisis period that the State formalized urban planning at the national level. One of the several amendments of Article 27 reinforced the decision-making and planning roles of the State regarding the fate of land use, natural resources, and human settlements.[81] Within this context, new planning institutions were created. For example, *Planeación del Desarrollo* or Development Planning was an emerging planning institution that produced the National Urban Development Plan. In addition, the *Secretaría de Asentamientos Humanos y Obras Públicas* or Ministry of Human Settlements and Public Works was created to direct national urban policy.[82] These institutions were the seed that caused the decentralization process in Mexico to grow.

As part of that decentralizing movement, existing planning institutions, such as INDECO, were closed and others like CORETT underwent restructuring. Although CORETT continued to exist, it could no longer request *ejido* land for expropriation or plan space for future urbanization. Those functions were relegated to state governments. CORETT's only function remained that of regularizing informal settlements through the National Plan for Land Tenure Regulation implemented from 1978 to 1982.[83]

NEOLIBERAL URBAN AND REGIONAL POLICIES

The economic crisis, prolonged into the Miguel de la Madrid term (1982–1988), was managed by initiating privatization of State owned industries and by controlling wages. The economic strategy that had historically benefited investors was implemented again under the guise of the so-called "economic restructuring." The years of this *sexenio* were part of what was known as the lost decade because from 1982 to 1987 and beyond, Mexico attained zero economic growth. In addition, the effects of this recession were high rates of unemployment, low wages, and a decaying quality of life, especially for the working rural and urban populations.[84]

In contrast, export-led sectors such as the automobile, beer, cement, and glass industries, mostly located in Northern regions reported continuous growth. In the same way, the assembly line plants (*maquiladoras*) on the border area retained their vitality.[85] It might be speculated that the North did not lose its economic dynamism because of the strength of its economy versus that of the South. However, the regional policy of that period clearly favored the Metropolitan area of Mexico City, the border cities (*maquiladoras*, industrial parks, energy-related infrastructure), and the

Northern states, to the detriment of the Southern states that were assigned less or no public investment. The only Southern areas supported in the 1983–1988 regional plan were tourist and oil producing cities such as Acapulco and Poza Rica located in Guerrero and Veracruz, respectively.[86] Thus, in addition to the economic crisis, the lack of financial support affected regions in uneven ways, further defining their inequalities.

To emphasize that inequality in relation to the policy of devaluating the peso, Aguilar Camín and Meyer state:

> The reverse policy of aggressive undervaluation of the peso since 1983, rewarded, on the other hand, another type of sectoral concentration— exporters, *maquiladoras*, and the tourism industry. To give an idea of the volume of the transfers to those sectors, between 1986 and 1987 the Mexican exporters obtained, according to estimations by the French economist Maxime Durand, an additional profit of approximately $4 billion, almost half of what was needed to service the Mexican foreign debt.[87]

It was during this period that deconcentration policies were initiated with the tentative plan to redistribute population and economic activities to medium-sized cities (those comprised of 500,000 to one million inhabitants.) Cities with significant concentration in industry, agriculture, and tourism would be heavily supported. The program was named The Medium Cities Program and began by providing basic urban services and infrastructure to those cities.[88] A second step was the creation of *Reservas Territoriales* or Territorial Reserves. These reserves would be formed by expropriating *ejido* and communal lands that the State would keep for later urban development. In order to carry out the reserves project, old institutions changed names and became new planning institutions. In 1982, SAHOP, the Ministry in charge of Human Settlements created in the Echeverría sexenio became the SEDUE, *Secretaría de Desarrollo Urbano y Ecología* or Secretary of Urban Development and Ecology. The new SEDUE had additional functions such as being able to act as "applicant and beneficiary of *ejido* expropriations and to constitute them into Territorial Reserves."[89]

Also, as part of decentralization, in 1983 SEDUE coordinated a national system of lands and conferred on the states the responsibility of applying their own urban land programs and of creating their own Territorial Reserves. As urbanization was reaching adjacent areas in some states, the law established regional urban planning cooperation.[90] The role of the State in urban land use planning and administration became stronger than ever.

TERRITORIAL RESERVES

The formation of the Territorial Reserves was not free from opposition. Affected *ejidatarios* who either did not want their lands to be expropriated or wanted to receive a fair payment for their lands, were against the project of Territorial Reserves. Since the 1940s expropriated *ejido* and communal lands had been paid at their agricultural instead of their commercial values. *Ejidatarios* rightly feared that they would be paid minimal prices for their lands as had always been the case.

By 1985, the opposition to the Territorial Reserves program prompted several changes in its implementation. The State raised the payment of the agricultural value for expropriated *ejido* land, although it still did not compensate expropriatees for the true commercial value of their lands. The second modification was to request that local governments, prior to expropriation action, sign agreements with *ejidatarios* to request their consent. This was a basic informational step that did not exist in the initial program. Local governments, however, continued to be the entity that unilaterally set the schedule of payments and the amounts to be paid for *ejido* land.[91]

By the end of the presidential term of de la Madrid, the agrarian and urban bureaucratic institutions competed for decision making power over the fate of *ejido* lands and the way they would be transformed. As part of that struggle, in 1988, the peasant national organization CNC or *Confederación Nacional Campesina*, proposed to organize, along with rural populations, real estate *ejido* enterprises. A prompt response to that proposal came from the de la Madrid administration under the form of a new bureaucratic entity: the InterSecretary Commission of Territorial Reserves and Regularization of Land Tenure. Three clashing Ministries participated in this commission: The Secretary of Urban Development and Ecology (SEDUE), the Agrarian Reform Secretary (SRA), and the Programming and Budget Secretary (SPP). These ministries represented different actors and interests: the SRA represented the rural sector, SEDUE embodied the urban sector, and the SPP held the resources available to develop either sector. Since the 1940s each sector had tried to obtain more control over decision-making on *ejido* lands and now they were part of a single entity.

Through the Inter-Secretary Commission, local governments were awarded control of *ejido* land through the creation of State enterprises that supposedly would include the *ejido* populations affected by expropriatory decrees.[92] The creation and new functions of the Inter-Secretary Commission led to the decentralization of CORETT to the different State governments. This movement restructured not only institutions and their functions, but mainly *ejido* land near major cities, the potential location of economic activities, and, consequently, the organization of regions and labor.

AGRARIAN VS. URBAN REFORMS IN THE PRIVATE INTEREST

At this point, institutions, policies, and actors made evident two contradictory and complementary processes regarding land in Mexico. On the one hand, through Agrarian Reform agricultural land could be expropriated from private hands in order to redistribute it in the form of *ejidos* to landless rural populations. On the other hand, through regularization processes,

ejido lands could be expropriated from *ejido* communities in order to redistribute them to the new urban populations. Apparently, while land for
Agrarian Reform was diminishing, the land for urban reform was increasing. It is necessary, however, to ask who the beneficiaries were and who the
losers were in these simultaneous processes. It is important also to note that
private land was unaffected by these processes. Private interests greatly benefited from *ejido* land conversion, but they were careful to implement different "exceptions" to Article 27 in order to protect their private lands.

Private land has rarely been expropriated for purposes of the common
interest. The *amparo* mechanism implemented during the Alemán tenure
guaranteed the protection of private lands. The *amparo* has become a legal
mechanism craftily manipulated by private landlords to preserve their
properties. *Ejido* lands, in contrast, have been extensively expropriated all
over Mexico with or without an *amparo*.

Private interests are not openly involved in the process of agrarian and
urban reform, but they are, in some magic way, found everywhere. They
have profited from expropriated *ejido* lands and have developed them also
illegally into profitable, high-income residences or space for the location of
different industries such as tourism and manufacturing. As the Mexican
economy was heavily nationalized, private interests became part of that
State and, consequently, had a final say in decisions affecting Agrarian Reform and Urban Reform. While executing urban and regional policies,
many public officials became owners of *ejido* land that they advanta-
geously acquired during the term of their administrative positions.

AUTHORITARIAN PLANNING:
TERRITORIAL RESERVES IN THE 1990S

In the 1990s, the State revived the 1980s Territorial Reserves initiative
with the purpose of providing land for housing and for investment space
for the national and international capital looking to relocate in medium-
sized cities. Decentralization was one of the key policies of this period and,
in order to achieve it, the national planning institutions designed the 100
Cities Program, which was a continuation of the Medium Cities Program
of the de la Madrid administration. As part of the 100 Cities Program, the
State enacted into Agrarian Law the expropriation of *ejido* land to "create
and expand Territorial Reserves as well as urban development areas, housing, industry and tourism."[93]

Another objective of the Territorial Reserves program was to harmonize
land use planning and management within a scope similar to that of a
Comprehensive Plan. Its goal, as stated in official documents, was to con-
template the participation of public and private entities by "linking urban
development with social development."[94] Typically, however, although the
law included the appropriate words, its implementation would follow a
top-down approach characteristic of authoritarian planning.

As designed, the Territorial Reserves program would be income selective—it would incorporate land for urban purposes and provide urban infrastructure only to those populations who fell into the income bracket of 2.5 times the minimum wage.[95] After decades of decreasingly lower wages, low-income people did not meet the income requirements.

The Territorial Reserves program was projected to obtain 125,000 hectares of land by year 2000. It was planned that 65 percent of those lands would be captured either from already privatized *ejidos* and rural communities, or from privatizing the share of common-use *ejido* lands that could later be transferred to a mercantile or private society.[96] The Agrarian Law specifies that common-use *ejido* lands are an economic reserve for the *ejido*. They cannot be sold or forfeited, except when the *ejido* assembly decides to do so.[97] The process of converting common-use *ejido* lands could work smoothly or could bring to light the internal contradictions and antagonisms within the *ejido* itself. As Territorial Reserves require more *ejido* land for urbanization or other purposes, more examples will shed light on the effects and implementation of this relatively new legislation.

The remaining 35 percent of the land that would become Territorial Reserves would come from the privatization of public lands (State lands), land donations, *ejido* real estate enterprises of private character, and through new expropriation of *ejido* and communal lands. In none of the mechanisms legally specified to procure land for the Territorial Reserves is the role of non *ejido* private land mentioned. It is evident that the participation of private land is not to be considered in the process of developing the Territorial Reserves. Thus, legislation on Territorial Reserves and its application will place the responsibility for housing people and industries fully on the shoulders of *ejidatarios* and their lands. It is obvious that *ejidatarios* did not design the 100 Cities Program or the tenets of Territorial Reserves of the 1990s. The conflicting agendas related to land and land use continue, now in the urban arena.

PLANNING INSTITUTIONS OR PRIVATIZING INSTITUTIONS?

In 1992, during the Salinas administration, SEDUE became SEDESOL, *Secretaría de Desarrollo Social* or Secretary of Social Development. SEDESOL would be in charge of regional and urban development policies and programs. Its jurisdiction extended to the regulation of housing and environmental programs. Along with this institutional change, the 1976 General Law of Human Settlements was modified in 1993. The new law emphasized the role of local governments in urban planning, land use, zoning regulations, and Territorial Reserves. In addition, it indicated that "transparency"[98] and social participation should be elements in the urban development process.[99] The law appeared to be a very progressive piece of legislation that tried to combine and harmonize different government sectors, planning institutions, and actors both in the rural and urban spheres.

It even highlighted its harmonization with the 1992 Article 27 regarding the use of *ejido* land for urban development and housing purposes.[100]

To achieve that harmonization, new institutions emerged to apply the 1992 Article 27 described in Chapter Two. The *Procuraduría Agraria* or Agrarian Attorney was the institution in charge of implementing the *Programa de Certificación de Derechos Ejidales y Titulación de Solares Urbanos* or Program of Certification and Titling of *Ejido* Land and Urban Lots (PROCEDE). Through this program *ejido* and communal lands would be measured and titled, both prerequisites for privatization. It was claimed that participation in the program was voluntary. Since the measurement of parcels renewed old clashes over *ejido* limits, the *Tribunales Agrarios* or Agrarian Tribunals would decide, by mediating and negotiating with the affected parties, on the limits of the *ejido*. Once measured, the plats[101] were recorded in the *Registro Agrario Nacional* or National Agrarian Records. This latter institution would issue, based on the plats, the certificates and titles to *ejidatarios*. Certificates were extended to those wanting to continue as *ejidatarios*, while titles were issued to those requesting privatization of their parcel. The new institutions had managed to generate and update property records and their land uses, which were fundamental activities to carry out planning. The same institutions provided new certificates and titles of *ejido* land that were required by the 1992 Article 27 for their further legal privatization. Those planning institutions were easing the process of future privatization of *ejido* and communal lands.

At the time those institutions were created, there were "29,000 *ejidos* and agrarian communities, which in turn incorporated 3.5 millions of *ejidatarios* and *comuneros*.[102] They owned approximately 4.6 million parcels and 4.3 million urban plots. The extension of *ejidos* represented nearly 50 per cent of the total national territory. The population that inhabited those spaces was over 25 percent of the total population of Mexico."[103]

CONCLUSIONS

As world conditions and immigration changed, and investments were restructured, the Mexican economy transformed *ejido* land, and new actors appeared on scene. Those actors had held different identities in different spaces and times. The rural and Indigenous populations, owners or not of *ejidos*, once displaced, went to the cities where most of them became part of the urban labor force living on the periphery of the city. Migration was not the only direction taken by rural populations. Those rural communities which were not displaced stayed in their *ejidos* and began renting them, while they combined their agricultural labor with other income-generating activities.

In the same way, not all the big rural landowners saw their lands affected due to the Agrarian Reform in Mexico. Most of them managed to continue their influence on large extensions of rural land by crafting exceptions to

Article 27 and, in this way, avoid the redistribution of their land monopolies. Others started speculating with land in the city. Over time their influence might have changed of location, but their speculative role continued either in the rural or in urban spaces.

As *ejido* land was taken to fulfill the needs of new immigrant populations in the city, as well as the needs for space on which to locate different economic activities, land use conversions took place thus speeding the process of the transformation of rural *ejido* land into an "urbanized *ejido*." *Ejido* land can be seen as the unit where different internal and external processes in Mexico took place. By external processes, I refer to the role of changing world conditions, and specifically the U.S. economy. This argument will be further developed in the next chapter. In any case, Mexico's economy swings back and forth as the U.S. economy fluctuates. The *ejido*, as an economic unit, is not free from the sway of national and international economic conditions.

The internal processes, on the other hand, are related to the regional and urban policies designed to direct the Mexican economy. Those policies restructured economic spaces and relations which, in turn, changed the allocation of investment, infrastructure, and the redistribution of land for certain economic functions. This chapter has laid out the development of urban and regional policies as a way to trace and reveal the urban-rural transformation of the *ejido*.

As *ejidos* were transformed so were their economies and populations. Thus, another task of this chapter was to make clear that *ejido* land use changes shaped the emergence of Mexican cities and regions. A constant interaction among *ejido*-city-region-city-*ejido* has been taking place in several directions. The *ejido* population going to the city transforms both the *ejido* and the city. Since these two spaces are part of a region, their transformation changes, in turn, the dynamics of the region where the *ejido* and the city are inserted. Likewise, capital investment in the region affects both the city and the *ejido*. The transfer of physical space between the city and the *ejido*, as part of the rural-urban transformation, shapes and reshapes the *ejido*, the city, and the region.

The different waves of economic restructuring in Mexico have created different geographies, predominant among them are Mexico City as a city-region, and selective areas such as the border, the tourism, and the oil-producing regions. Throughout recent history, Mexico City and the Northern regions have enjoyed public investment and the development of infrastructure. These two elements have accumulated and contributed to the current shaping of the regions.[104] They have also contributed to the uneven development of regions, with the South being the most disadvantaged.

Since the early migration of rural populations into the city, the separation between legal and illegal spaces has been evident. This separation is best described by Nicolás López Tamayo in his study of the urbanizing *ejidos* in the city of Puebla:

Urbanization in Puebla, in the last two decades, has been the result of the
articulation between legal and illegal land markets. Both markets are part
of a single process of capital accumulation taking place within the same
territory. As a result, two models of the city have emerged: the legal and
the illegal one. The high- and medium- income populations are the
authors of the first process. The low-income populations are the actors
that share the informal and illegal model of the city.[105]

As the State witnessed and promoted these two models of city, they subsidized capital. Capital displaced populations here and there, but it did not
contribute to their relocation. It was the State with its housing and land use
policies, that attempted to ameliorate the crisis for shelter that low income
populations have historically faced.

In the case presented by Castells, the fact that some settlers decided not
to legalize their lands was symptomatic of the extent to which land was so
costly. They could not afford legality.

The privatization of *ejido* land, brought about by the 1992 amendments
of Article 27, would send into "illegality" a new wave of dispossessed and
displaced *ejidatarios* and Indigenous communities. Nonetheless, privatization continues being the buzz word of economic restructuring all over the
world. The 1992 Article 27 is just a sample of a major process of economic
restructuring taking place worldwide, a process in which, through legislation and the modification of the functions of State and planning institutions, resources are being reallocated. In the specific case of Article 27, its
new legislation permits the redistribution of land to non rural and non
Indigenous populations. The new Article 27 signals the emergence of new
institutions that are redistributing the land required to settle economic
activities and labor in this "globalized" era. Chapter 4 develops in more
detail the character and meaning of the "new" State, the international and
national contexts of widespread privatization, and the global supranational institutions that have urged such a global redistribution and privatization of space.

Chapter Four

Privatization of *Ejido* Land in the Age of NAFTA

The prior chapters have provided a context for understanding the urban and regional policies that permitted land privatization as well as the role of Article 27 and the *ejido* system in the regional, urban, and rural land use patterns, that is, in the shaping of cities and regions in Mexico. In contrast, the central objective of this chapter is to lay out the roles of supranational and national institutions—the World Bank, the International Monetary Fund, private international and national banks, and the Mexican State—and their policies that weaved into the global context that instigated the transformation of Article 27 and the privatization and commoditization of *ejido* lands in Mexico in the age of free trade.

At the global level, the policy of privatizing *ejido* land is the result of a process in which at least two main external and extra-national pressures took place. First, the transformation of land policy, represented by Article 27 of the Mexican Constitution, that supranational lending institutions demanded from Mexico in order to make its economy more efficient and capable of paying back its foreign debt. Second, the influence of global financial capital that, through the North American Free Trade Agreement (NAFTA), contributed to the privatization and subsequent commoditization of multiple economic sectors, including land.

At the national level, the Mexican State has consented to the gradual implementation of economic restructuring programs, also called the neoliberal model. Neoliberalism has invoked the integration of Mexico into the global economy by transforming the functions of the national State and its relationships with different economic and political actors. This transformation required a "retreat" of the State from participation in the economy through the widespread privatization of various economic sectors. In the specific case of Article 27, its transformation represents the legalization of the privatization of *ejidos* and the breaking of the social contract embodied in Article 27 and the *ejido* system. The 1992 Article 27 allows the legal conversion of *ejido* land into a commodity. According to the Mexican privatizing State of the NAFTA era, privatization equates "modernization." The development discourse used by the Mexican State traditionally looked upon *ejidos* as a backward and inefficient form of land tenure in need of "modernization." Communal and *ejido* lands have always been associated, especially in the Central and Southern regions of Mexico, with Indigenous communities and, consequently, with all the "vices" of underdevelopment, namely being rural, nonmechanized, and subsistence-oriented.

Before Article 27 could be modified by the Mexican government, the State also had to modify its own functions and relationships in order to accomplish privatizations. The neoliberal Mexican State of the 1990s had to use a deceptive discourse to cover the privatization of land and to exalt the integration of Mexicans in the First World.

This chapter highlights the role that Mexico's foreign debt played in the configuration of the global context that pushed the implementation of economic restructuring programs and, consequently, of privatizations. The privatization of *ejido* lands at the global level opened the integration of those lands not only to the national, but to global market. Global capital was in need of new spaces of investment and privatization created those spaces by "liberating" previously regulated territory, the *ejidos*. As supranational and national financial institutions, in conjunction with the Mexican State, are the entities directing this commoditization and the creation of a new geography of investment spaces, this chapter claims that globalization is just another word for imperialism, a way of redistributing the global territory for the survival and expansion of global capital.

WHAT DOES DEBT HAVE TO DO WITH THE PRIVATIZATION OF *EJIDO* LAND?

The intention of this question is to highlight the relationship between the supranational financial institutions, the Mexican debt, the transformation of Article 27, and the "deregulation" of *ejido* lands. The answer has to do with a local-national-global interaction. This interaction is illustrated by the transformation of Article 27 of the 1917 Constitution, and consequently of land policy in Mexico, which was one of the various requirements established in the structural adjustment programs designed by financial global institutions—the World Bank, the IMF, and private banks—in order to transform Mexico's economy and enable it to pay and renegotiate its debt, and to continue being eligible to apply for loans.[1]

Privatization and "clarification of property rights"—for the private sector—were two of the main elements that commanded the structural adjustment programs.[2] This clarification of property rights in fact existed in the 1917 Article 27. As Chapter Two points out, the 1917 Article 27 regulated land and natural resources in Mexico and included private, public, and social property. The Mexican Constitution considered communal and *ejido* lands as social property and had planned its protection by stating in Article 27 that those lands were not legally alienable, meaning they could not become a commodity. Chapter Three explains how the *ejido* system functioned as an unintended growth control mechanism. It states that legislative protection and growth-control mechanisms were not compatible with the "free investment" tenets of the era of NAFTA. Consequently, as a way to secure their future loans and investments, the national and global private financial sector required that the Mexican State legalize the sale of *ejido* land, which would become their guarantee in case they needed to confiscate it. The collusion of those financial institutions and the Mexican government changed the Mexican Constitution to legalize the conversion of communal and *ejido* lands into private property. This was the "clarification of property rights" that the national and global capitals demanded. Within this context a second question arises: how was that debt created in the first place?

THE DEBT CREATED BY THE IMPORT SUBSTITUTION MODEL

Previous to the application of the restructuring adjustment programs of the late 1980s and early 1990s, Mexico, as well as many other Latin American countries, had followed an Import Substitution (ISI) model of economic development since approximately the 1940s. The ISI model had advocated for "programs of domestic industrialization that would allow countries to manufacture at home goods that were previously imported."[3] The major assumption underlying this model of development considered industrialization, as opposed to agriculture, as the engine for economic growth. Another assumption was that, by producing their own goods, so-called developing countries would gradually decrease their dependency on developed countries. Paradoxically, the ISI program drove Mexico, as well as other Latin American countries, into a new dependency cycle as it constantly needed machinery from developed countries and more capital from international financial institutions in order to sustain the ISI development model. In addition, during that period, investment focused on industrialization and commercial agriculture which negatively impacted *ejidos* and communal lands.

At some point, the Mexican government could not possibly continue paying the debt acquired through years of borrowing from financial lending institutions. Given the 1982 oil crisis that seriously affected the petrolized Mexican economy, plus the social programs that Mexico sustained with low-interest loans borrowed from the very same international financial institutions that would eventually restructure its economy, Mexico faced a huge debt crisis and declared that same year a moratorium on debt payments.[4]

RENEGOTIATING THE DEBT PAYMENTS

Mexico, in conjunction with other Latin American countries, might have had the option of forming a continental front to renegotiate their debts as a bloc, instead of individually, as they chose to do. A bloc of Latin American countries could have strengthened their position at the negotiation table with global financial institutions. However, in order to understand the coercive character of such a negotiation, it is important to mention that the supranational lending institutions, namely the World Bank, the IMF, and private banks, unilaterally prescribed a debt renegotiation on a country by country basis.

Those financial institutions not only controlled the renegotiation process, but also manipulated its outcome by conditioning the lending of additional funding based on the acceptance of the economic restructuring program.[5] Those negotiations highlighted the dependency of Mexico and other Latin American countries on outside money, as well as the power of financial pressure exercised by supranational institutions that could, and still can, impose policies and modify national legislation and planning in connivance with national economic and political elites. The negotiations also showed the nature of the national private economic elites that supported such a process. Those elites resembled more the legacy of the Northern agenda of the Mexican Revolution, an agenda that protected and promoted private property interests to the detriment of the South.[6] This collusion between national economic elites and supra-national financial institutions took place not only in Mexico, but all over Latin America, and had a disadvantageous impact on the renegotiation of the national debt at the continental level.

In their study of the debt crisis in Latin America, Griffit-Jones and Sunkel criticize the coercive tactics and programs of those international financial institutions and state that continuous transfers, from poor to rich countries, in the form of debt interest and payments, had surpassed the amount of total loans initially awarded by those financial institutions. They also affirm that, after a decade of transferring net resources from debtor countries to creditor banks, those financial institutions had long recovered from the outcomes of the 1982 debt crisis, to the point that they had recouped all their losses. The same statement, however, could not be made by the debtor countries that faced the imposition of the economic restructuring programs.[7] Reynolds summarized that situation by emphasizing that those debt payments "became the basis of an unprecedented net transfer of Latin American real resources to industrialized countries."[8] In short, the renegotiation of debt payments revealed the global financial institutions as modern usurers that increasingly profited from the financial traps in which Mexico and other Latin American countries had become enmeshed. This was the context in which economic restructuring programs took place.

THE ECONOMIC RESTRUCTURING
PROGRAMS OR NEOLIBERAL MODEL

The principal objective of economic restructuring programs was to prepare Mexico and other Latin American countries to pay their foreign debts by changing the way in which their economies were operating. It was the opinion of the global financial institutions that the model of development followed by "underdeveloped" countries was not appropriate. They argued that economic growth was not taking place which, in turn, affected in a negative way the profitability and efficiency of the Latin American economies and their capacity to repay their debts.[9]

Gurría pinpointed August 22, 1982 as the date when Mexico applied for a three month extension to resume payments of its debt, which, due to the increase in interest rates and combined with the fall of oil prices, had become unpayable. In 1983, Mexico renegotiated alternatives to fulfill its debt payments, which meant the acceptance of the application of an economic restructuring program or neoliberal model.[10]

Contrary to the ISI model, the neoliberal model promoted export-oriented industrialization, opening of the economy to foreign investment and trade, elimination of public subsidies, wage controls, legislative changes to accommodate private investment and, of course, privatizations.[11]

Global and national financial institutions directly intervened in the design of the policies that would restructure the Mexican debt.[12] This intervention would modify the regulation and operation of the Mexican economy in favor of the private sector. Since Article 27 of the 1917 Constitution regulated the property of land and its resources, which were essential elements of the Mexican economy, they had to be transformed, too, especially those related to *ejido* and communal lands because they were the land tenure property systems that were not private nor legally permitted to be privatized. Among the global financial institutions intervening in this policy design were the IMF, the World Bank, and private banks.

THE ROLE OF THE IMF, THE WORLD BANK, AND PRIVATE BANKS

The presence of international financial institutions or supranational lending institutions has been constant in Mexico's life since the declaration of the 1982 debt crisis. Their intervention has ranged from the conditioning of future loans on the basis of the application of specific policies, to the monitoring and supervision of the restructuring adjustment programs and policies, by requesting timely economic reports from Mexico. They have also advised Mexican technocrats on strategies to undertake privatizations and have even "creatively" devised a "debt-securitization" program that "called for the conversion of loans into tradable bond instruments."[13]

In addition, these institutions have "suggested" export production policies, the reduction of the governmental sector—its functions, operations, legislation and regulations—and, consequently, the privatization of State companies.[14] Finally, they have supported and advocated free trade. In short, their role has been to completely change the nature of the Mexican State and to create favorable conditions for the relocation of global private investment.

Throughout this process, these institutions have agreed to award Mexico the necessary loans, as long as those funds were used to increase efficiency and productivity, which meant the use of loans to support the policies of structural adjustment. This loan process became a self-renewing cycle as new loans were required in order to continue paying the debt. It seemed that the main role of these institutions was to preserve and prolong the lending cycle. Like the *hacienda* store of the Porfirio Díaz era described in Chapter Two, Mexico is continuously indebted, for generations to come, and no matter what Mexicans do, they will forever be paying the acquired foreign debt. This process resembles the one described by Lenin:

> Finance capital, concentrated in a few hands and exercising a virtual monopoly, exacts enormous and ever-increasing profits from the floating of companies, issue of stock, state loans, etc., tighten the grip of financial oligarchies and levies tribute upon the whole of society for the benefit of monopolists.[15]

Although time, circumstances, and place have changed, the process seems to repeat itself. The era of free trade, of NAFTA, is the current time. One the one hand, the indebted countries and, on the other, the search by global capital for new spaces of investment, compose the current circumstances. The policy connecting the current time, circumstances, and place are the structural adjustment programs.

THE 1985 AND 1988 STRUCTURAL ADJUSTMENT PROGRAMS

The specific outcomes of the intervention of international financial institutions were the Baker and the Brady Plans, which were issued in 1985 and 1988, respectively. Both plans were implemented in Mexico and they marked the beginning of the major structural adjustment programs in Mexico. The Baker Plan called for the Mexican State to open local markets to imported products by reducing tariffs for foreign products; to prepare favorable conditions for foreign investment by changing the legislation that regulated it; to initiate a privatization program of nationalized key economic sectors; and to reduce its public expenditure.[16] Later, the Brady Plan focused on expanding the privatization program, promoting export-oriented production, securing legal and infrastructural conditions for foreign private investment, and on liberalizing trade.[17] The Baker and the Brady structural adjustment programs called for a departure from the nationalist and inward oriented economy that the Mexican State had implemented in various ways since the end of the 1910 Revolution and later through the ISI model. These plans did not consider, however, that the "nationalist" economy was a term that included conflicting agendas at the local, regional, and national levels. As Chapters Two and Three have explained, different ethnic groups, classes, cultures, and regions have constituted the internal history of a Mexico. Some of the country's elite groups had gradually pushed forward several "exceptions" in the "nationalist" agenda of the Mexican economy and specifically in the land legislation-Article 27. Those groups had been preparing the ground for making those exceptions the rule.

PREPARING THE GROUND FOR
FOREIGN INVESTMENT IN MEXICO

The first change made by the Salinas' administration, to legally accommodate the possible investment outcomes of the restructuring adjustment programs and later of NAFTA, was the 1989 Regulations of The Law to Promote Mexican Investment and to Regulate Foreign Investment, which replaced the 1973 Law of the same name.

In title, both laws remained identical, except for the year in which they were issued, but, in 1989, the content of the 1973 Investment Law was radically modified. Before 1989, the investment legislation restricted the proportion of foreign capital and ownership. Some economic activities were exclusively reserved to the State, and some others were limited to operation by nationals of the country. Most of those nationals were the new generations of *Criollos*. Under the new regulations, Mexico began privatizing State businesses, deregulating foreign and national investments, and providing significant incentives and subsidies to foster foreign investment. As stated in the new law: ". . .Mexico has initiated an opening of its economy in order to participate successfully in international trade and investment flows."[18]

The Foreign Investment Law of 1973 set a limit of 49 percent foreign ownership for any company operating in Mexico. This law also restricted the acquisition of shares by foreign capital. In contrast, the 1989 Investment Regulations modified the percentage in foreign ownership and eliminated the percentage of shares of foreign capital. Currently, foreign investors can participate at any level, even at 100 percent, in the capital stock of most enterprises, without having, as before, to request the Foreign Affairs Secretary's authorization to do so.[19]

The law that regulated the repatriation of profits and dividends was also modified. With the new regulations, all profits and dividends can leave Mexico in their totality and free of taxes. In addition, the Salinas administration arranged legislation to protect the "intellectual property rights of large transnationals (mostly based in the U.S.)," while clearing up similar legislation for national intellectual property rights.[20]

By removing the limitations to foreign and national investments, the Mexican State of the 1990s ignored the history that prompted previous investment laws and the limitations to foreign capital. At the same time, the State showed its authoritarian character since in Mexico, laws, planning, programs, and policies have historically been approved without the consent or the knowledge of the communities affected. The removal of those limitations permitted global capital to relocate in Mexico. Some of the economic spaces that global capital would occupy would be the ones left vacant by the "retreat" of the State. Privatizations in Mexico had created new economic spaces for global capital.

THE PURPOSE OF PRIVATIZATIONS IN MEXICO

Most of the literature about privatizations in Mexico has focused on the enterprises owned and/or managed by the State, but not on land and natural resources. The rationale detailed in that literature to justify the purpose of privatizations was to "modernize" the economy by reducing State intervention.[21] The modernization project—the neoliberal model—would attack the high costs of production, the inefficient and bureaucratized labor, the waste of resources, the low productivity, and the lack of capital investment, among other elements.[22] Similar arguments were stated as the basis for modifying Article 27 of the 1917 Constitution: to modernize the agricultural sector, to reduce state subsidies, to make more efficient the countryside by allowing *ejidatarios* to first become private proprietors and then by being able to establish joint ventures with private capital.[23]

Although the purpose of privatizations was to redefine the Mexican State and transform the Mexican economy, for Mexican capitalists, privatization represented the opportunity to expand and create partnerships with foreign investors, especially from the United States.[24] This partnership would secure financial and human resources necessary for their own expansion, Concheiro described this form of partnership as "strategic alliances."[25] Evidently, in the case of the countryside, not all *ejidatarios* would be able to establish the same kind of alliances. Regional, economic, ethnic, and cultural differences will influence, and probably preclude, the establishment of those alliances.

When evaluating the purpose of those privatizations, Aspe, one of the architects of the privatization process in Mexico, acknowledged that it was not possible to define which form of property, public or private, was "superior." Consequently, he wisely advised that the decision to privatize would depend on the specific characteristics of each industry, enterprise, or sector.[26] This statement was also voiced by other policy makers, since, in fact, there was no evidence that privatization would lower costs of production, make labor efficient, increase productivity, and attract capital.[27]

Ramamurti had also warned that "it would be naive to expect the change in ownership alone to deliver the efficiency gains expected by policy makers."[28] He also cautioned about the possible "unintended problems down the road if [privatization were] not [going to be] used carefully."[29] Nonetheless, privatization processes took place indiscriminately in all sectors on which the State had had some economic or planning intervention, including land and natural resources. The widespread adoption of this policy reflects the character of the elites within the Mexican State. These elites have not changed substantially, either in origin or in intention, since colonial times. Historically, they have always privatized Indigenous and communal lands. The tacit purpose, then, of implementing the privatization of *ejido* lands is to devise another attempt to facilitate privatizing the lands of Indigenous communities and that of other rural populations, but also to create new spaces of investment for both the national and global capitals.

THE ROLE OF THE MEXICAN STATE
IN THE PRIVATIZATION PROCESS

As mentioned before, at the international level, the 1980s' privatizations in Mexico were part of the structural adjustment programs that supranational financial institutions demanded of debtor countries if they wished to keep having access to further loans.[30] Although those institutions urged and put pressure on Mexico to carry out restructuring programs and to organize "property rights to include the private sector";[31] it is necessary to acknowledge that the national political, economic, and planning elites consented to such a transformation. Weintraub equates this consenting attitude as a "remarkable change of thinking from what existed in the LAC [Latin American countries] region then to what prevails now."[32]

According to Ross Schneider, the national State could covertly use privatization to court the private sector, to discipline labor, and openly to "improve the investment climate."[33] When reporting the results of an interview with Jesús Silva Herzog, the then Minister of Finance, Ross Schneider indicated that the purpose, at the domestic level, of bringing about privatizations changed during the six years of the de la Madrid administration.[34] According to Ross Schneider, Silva Herzog classified those changing purposes in four chronological stages.[35]

THE PURPOSES OF THE FOUR STAGES OF THE PRIVATIZATION PROGRAM

During the first stage of the privatization program—around 1983–1984—the de la Madrid *sexenio* concentrated on healing the hurt pockets, and gaining "the confidence" of private investors who had "suffered during" the Echeverría[36] and López Portillo[37] regimes. As elaborated in Chapter Three, during the Echeverría administration (1970–1976), land had been expropriated extensively in order to fulfill demands of land redistribution; meanwhile, during the López-Portillo period (1976–1982), the banking system had been nationalized.[38]

In a second stage, between 1985 through 1988, the privatization program aimed at "reducing the deficit and administrative chaos" that dominated the Mexican State bureaucracies. The publicly stated objective of this second stage was the reduction of the participation of the State in the ownership, management, and entrepreneurship of the national economy.

According to Silva Herzog, the presidential electoral politics in 1988 gave character to the third stage of the privatization program. During that period, the objective of the program was to appeal to the right wing partisans sympathetic to the *Partido Acción National* or National Action Party (PAN), many of whom belonged to the Mexican business elite.[39] This move attempted to lessen the increasing weight of opposition parties that threatened the influence of the PRI, which was the "official" party.[40] At that time, the PAN was considered the second political force in Mexico.

Finally, during the fourth stage of privatizations, the de la Madrid government (1982–1988) undertook the dirty work of quickly privatizing some State firms, so that the incoming president, Salinas (1988–1994), would not inherit that function and start his term with the stigma of those initial privatizations.[41]

PERIODIZATION OF THE PRIVATIZATION PROGRAM BY SECTORS

While Silva Herzog described a periodization by objectives, Ramírez reported a periodization by economic sectors of the privatization process elaborated by the *Secretaría de la Contraloría General de la Federación* or General Comptroller of the Federal Government.[42] This periodization identified three phases of privatizations and the sectors affected in each phase. During the first phase, December 1982 through January 1985, "nonpriority state-owned enterprises in the manufacturing, textiles, and hard consumer goods" were sold, liquidated, transferred, or merged. The Monterrey steel mill, Mexicana Airlines, and a hotel corporation were the most significant privatizations during the second phase that lasted from February 1985 through November 1987. The last and third phase of privatizations, December of 1987 through November of 1988, witnessed the most intense period of privatizations in Mexico. During that period, sectors that were previously considered strategic such as mining, auto parts, fertilizers, and sugar started to be privatized.[43]

The stages proposed by Ramírez concentrated on the privatizations undertaken during the de la Madrid administration in order to speed the privatization process. Ramírez states that at the beginning of 1982 there were more than 1,155 public entities, but by the end of the de la Madrid *sexenio*, in 1988, that number had decreased to 711, which meant that about 62 percent of those publicly owned entities had been privatized.[44] The transformation of the State was under way and this was reflected in the decreasing land redistribution that culminated in the transformation of Article 27 during the Salinas administration.

PRIVATIZATIONS UNDER THE SALINAS ADMINISTRATION

The most relevant privatizations took place under the Salinas administration.[45] Those privatizations represented accurately the shock therapy advocated by the World Bank and IMF entities: they were quick, radical and, according to economists from those institutions, irreversible.[46] By May 1992, Salinas had privatized about 488 out of the 711 public entities he had inherited in his administration, which represented a privatization of 69 percent of the total State enterprises during the first four years of his administration.[47] Ramírez accurately described that, during the Salinas period, it was evident that "the pace of privatization ha[d] increased to the point where it [was] no longer a question of simply rationalizing public expenditures, but one of transforming in a fundamental fashion the structure and role of the State in the economy."[48]

The sectors affected during the Salinas *sexenio* were constitutionally considered strategic. In order to legally undertake those privatizations, Salinas changed the legislation, specifically Article 28, that assigned planning functions to the State as well as its exclusive control of "oil, basic petrochemicals, electric power, nuclear energy, satellite communications and railroads," to mention some of the sectors in question.[49] Among the sectors that were privatized after 1988 were the banking system, telecommunications, highways, and mining companies. The jobs lost due to privatizations, without considering their multiplying effects, were more than 200,000 between 1983 and 1989.[50]

It was in 1992, during the Salinas administration, that the transformation of the 1917 Article 27 or *Ejido* Reforms took place. Since land and natural resources were considered constitutionally as "property of the nation," the *Ejido* Reforms denationalized and privatized what previously was considered the national patrimony of Mexicans. Opponents to privatization pointed out that these processes could "lead to yet more concentration of wealth."[51] A high possibility existed that privatization of *ejidos* would facilitate the monopolization of land in private hands and the expulsion of the rural labor force, as *ejidatarios* in need could sell their land or use it as collateral for loans, and eventually lose it. Consequently, processes of land reconcentration and proletarianization of *ejido* labor force would take place. However, Salinas claimed that new opportunities would be open for the countryside as *ejidos* could enter into joint ventures with the private sector.[52] Evidently, privatization had a different meaning for different groups of Mexicans.

THE MEANING OF PRIVATIZATION

Since privatizations would alter what constitutionally was considered national patrimony, the word privatization was, at that time, avoided in official discourse. Instead, the 1983-1988 National Development Plan or *Plan Nacional de Desarrollo* referred to the privatization program as *desincorporación* or disincorporation.[53] The word disincorporation simply implied that a certain enterprise was going to stop being part of the State. Had the right word—privatization—been used, concerned groups might have opposed this program from its very onset. But the Mexican State curbed potential opposition to the privatization program through obscuring its real meaning and by using deceptive discourse.

Disincorporation was not the only word used to refer to privatization; other words used were divestiture, liquidation, transfer, sale, modernization, opening, liberalization, deregulation, thinning of the State, and streamlining of bureaucracy.[54] In the case of the 1992 Article 27, the words used were deregulation, titling program, and *Ejido* Reforms. These words did not convey the meaning of privatization as a policy that converted land into saleable merchandise, into a commodity. They did not convey either that privatization transferred ownership from public to private hands, from the nation to private investors, and from the State to the national and transnational banking system.

The Mexican technocrats might have avoided the word privatization because they were aware that, in the minds of the Mexican people, the word privatization communicated the opposite meaning to "national patrimony."

Privatization and "national patrimony" were words intimately related to property and to the recent and not-so-recent colonial history of Mexico. The land, the State enterprises, and other economic sectors that the "disincorporation" was going to affect had been, after the 1910 Revolution, expropriated from foreign and national private capital and became property of the nation. In any case, that private capital was identified with a history of injustice, inequality, and dispossession.

The property of such economic sectors as mining, land, oil, and electricity, among others, was not only relevant for the country's economy or the role of the State, as most literature on the topic has portrayed. Rather, this property was seen as symbolic of the sovereignty of the State, of nationalism, and of the nation understood not as the State, as in the Western thought, but rather as the Mexican people. As Erfani states:

> In the early postrevolutionary era, the political stability of the Mexican nation-state was culturally grounded in the modern state's symbolic links to Mexico's indigenous populace. In fact, the "revolutionary" state of the 1920's and 1930's came to symbolize the triumph of Mexico's indigenous civilizations over foreign intruders. The sovereignty of Mexico's revolutionary nation-state symbolized indigenous freedom from foreign domination and popular political self-determination.[55]

Economic and political elites constructed the idea of sovereignty on the basis of the Indigenous identity because, in spite of the *mestizaje*[56] in Mexico, most of those Mexican elites continue to be foreign, white, and segregated in their own privileged spaces, while the majority of Mexicans do not share the same characteristics. The majority of Mexicans have Indigenous roots.

Thus, the meaning of property and its rights has been intimately related to the cultural identity of Mexicans, shaped as a result of colonization, a war for independence, and the historical experience of one of the major agrarian revolutions of the 20th century, the 1910 Mexican Revolution. It has also formed a vision of the State as a genuine representative of the people's interests. The conceptualization of the nationalist and legitimate State had also been manipulated by the PRI since the 1940s. Fictitious land redistributions, invasions of *ejido* lands carried out by the PRI organizations, and repression of peasant and urban movements, managed to temporarily keep rural and urban populations from revolting. In the context of a declining legitimacy of the PRI, privatizations started affecting the property of goods that were previously considered beyond the realm of individual interests, as was the case of land and natural resources, both renewable and non-renewable. The Mexican State responded to the demands of becoming a new and privatizing State.

THE NEW ROLE OF THE PRIVATIZING STATE AND ITS NEW RELATIONS

Because the safeguard of property is undertaken by the State, the process of privatization is actually designed to shrink the role of State intervention, apparently one of the objectives of the restructuring programs. Paradoxically, as I have mentioned before, it was the State itself that was implementing, administering, and carrying out the privatization program. The State, through its different ministries enacted policies, regulations, and legislation to effectuate the liquidations, eliminations, transfers, or sales undertaken since 1983.

Ironically therefore, although the role of the State was supposed to be diminished, the implementation of the privatization program led to new administrative and bureaucratic functions. That also became the case with the transformation of Article 27 which implemented a titling and certification program of *ejido* lands and, in the process, created new State agencies to carry out those programs and to settle land boundary disputes.[57]

Through privatization processes, the State acquired the function of transferring resources from one group to another. A very pertinent question about this transferring of property rights is the one posed by Ramírez: ". . . at who's expense and for whose benefit?" If this transferring has reconcentrated resources in the hands of a reduced group of financial and economic interests, while "the majority of the Mexican people see their standard of living deteriorate daily,"[58] then it is necessary to expose such reconcentration processes and use accurate words to describe what privatization really is in Mexico: the reconcentration of wealth, the private monopolization of natural resources and land, and the transfer of property to the rich.

With regard to those changing State functions, Cassesse asserts that the privatization process has not meant a retreat of the State; on the contrary, it has meant a "reorganization" of its functions. Before privatization, "the state was an owner; after privatization, the state became a regulatory body of property."[59] In the specific case of Mexico, the 1917 Constitution did not describe the State as owner of land; rather it was the nation, which meant that the Mexican people were the proprietors of land and natural resources, not as individuals, but as a collective. The idea of property in Mexico and the idea of property renegotiated with international financial institutions reflect diametrically opposed experiences and histories. The Mexican experience and history refer to colonialism, racism, anti-imperialism, and an agrarian revolution. These topics were previously introduced in Chapters Two and Three of this book.

Finally, when trying to determine who in Mexico has undertaken the regulatory functions, and who has benefited from them, it is very difficult to separate different actors. Plenty of cases exist in which the political and economic power that a civil service position grants has been used to gain economic advantage. During the ruling times of the PRI, most of its high-level civil servants were, at the same time, owners of capital. There were no boundaries between policy makers, managers, and capitalists. They were one and the same. Therefore, if the State had created new rules to allocate property, this allocation was not free from conflict of interests, since the regulators were simultaneously investors. As a result, Ramírez foresaw an "unprecedented concentration of productive and financial resources in relatively few and powerful segments of the private sector (both domestic and foreign)."[60]

Privatization modifies, not only property and "the transferring of state owned properties to the private sector," but also, as Aspe affirms: "it implies the redefinition of the role of the State and of the civil society in the production process and in the income distribution."[61] Although many definitions exist to define civil society, it seems that Aspe includes, in his definition, both private and national investors. In other words, what Aspe tried to explain is that privatization of the property of the nation created a new relationship between the State and private interests which, in turn, changed the relationship of the State with other sectors of Mexican society. Or as Ramírez expressed: "the privatization of state firms in priority and strategic sectors goes beyond simply getting 'prices right' and increasing relative efficiency, it also has a political dimension which falls nothing short of undermining the relative autonomy of the state vis-à-vis powerful domestic and foreign interests (mostly U.S. in origin)."[62]

The discourse to define this relationship between the State, the Mexican people, and private interests has also radically changed. Before the application of economic restructuring policies, "the nation" referred to the Mexican people; meaning the working classes, the middle classes, and the peasantry.[63] Because of their history, Mexicans believed that private interests, either national or foreign, had not been favorable to the interests of the nation (in other words, themselves). The new discourse avoided the word "nation" and changed it to "civil society."

THE NEOLIBERAL DISCOURSE

Discourse was a powerful tool used to distort the reality of the privatization process. By dissecting the words of that discourse, it is possible to expose its intentions. To illustrate, here is a quote from the speech that Agustín Legorreta—a member of the well known aristocratic family of the Porfirio Díaz era, and during 1973, both director of the National Bank of Mexico and simultaneously president of the Association of Mexican Bankers—gave as part of his annual activities report for the Association:

> We, the State and the private sector, are, in the end, the same thing. It is a genuine deceit to present them not as distinctive elements, but as opposite parties. . . .[. . .]. . . We will never be able to divorce the interests of the State and the interests of the people. We are irrevocably united. Not private sector against the State; but private sector, all the people, with the State; and the State with all the people.[64]

Although that speech was given in 1973, it accurately reflected the nature of the "new" discourse in the 1980s, which suggested that the private sector, and specifically the banking system, had become the people.

Under such a self-denomination, they requested a "reprivatization"[65] process of the economic activities that they once owned during the Porfiriato era and before.[66] They wanted back the property that they had partially lost after the 1910 Revolution. In this scheme of things, the private sector was requesting that "the majority of economic activities now controlled by the State be put again in private hands, in that way, the government would become the efficient central institution: it would be the director and coordinator of the national economy; not the executor of all the tasks."[67] In conclusion, as Grosse states it: "the 'solutions'. . . . originated in the private sector."[68]

The new discourse assembled by the national private sector, instead of questioning the intervention of the Mexican State in the national economy, corroborated State intervention, but modified its role.[69] The new functions of the State would be to transform the legislative framework of the economy, to invalidate the history and the ideology that supported that framework, and to renegotiate the relations or alliances established after the 1910 Revolution. The new role of the State at the national level would be to "promote the conditions that allow the [participation of] private investment and [the guaranteeing of] its ample profits."[70] In Otero's words, "[y]et, the main implication of such reforms is to modernize authoritarianism, rather than transform it."[71]

EFFECTS OF THE ECONOMIC RESTRUCTURING PROGRAMS

Aspe had argued that the income generated by the sale of State enterprises would be directed to decrease the Mexican debt.[72] He imagined that this transaction would eventually save the resources that the State was using to pay for the debt. He believed that those savings could then be used to increase social expenditure in order to prevent the negative impact of the restructuring programs. His belief would be partially corroborated and partially challenged by reality. The income generated by privatization processes was used to pay the debts contracted with creditor banks; however, the savings that Aspe forecasted were not used to generate more jobs to pay for social programs. Had they been, the money would have helped compensate for the decrease in jobs and wages that this economic policy had generated. Thus, workers and other nonprivileged sectors of the Mexican society paid in full for the negative impact of the structural adjustment programs.[73]

According to Brailowsky, the result of the debt crisis proved to be very favorable for the creditor banks, however, for Mexico it meant the total disorganization of its economy.[74] The implementation of structural adjustment measures reduced the level of employment, decreased wages, weakened the existing industrial national base, and the amount of the external debt skyrocketed, which was precisely the problem that creditor banks supposedly wanted to solve.[75] By 1990, Mexico was the "world leader. . . in privatizations."[76]

MEXICO IN THE GLOBAL ECONOMY: THE CREATION OF NEW SPACES OF PRODUCTION AND CONSUMPTION

One of the first and assumed motives in carrying out the economic restructuring programs was to generate revenues that would allow debtor countries to continue paying off their debts. However, once in the process of implementing those programs, the private sectors of developed and developing countries promoted their own interest in securing trade agreements. On the one hand, the developed countries had an interest in exporting capital in order to continue accumulating capital; on the other, the economic elites of developing countries wanted to secure "their insertion into the world economy, while taking advantage of regional markets."[77]

The United States and Canada were undergoing a different crisis, not debt-related, but profitability- and market-related. Their products were more expensive to produce than those from Japan and Europe, their competitors. For the United States and Canada, Mexico represented a new place where they could reduce their costs of production and increase their profits. Mexico promised abundant and cheap land and labor force, and a lax regulatory framework for industrial and agricultural production. In those cases in which earlier legislation had posed an obstacle for investment, it was rapidly included in the restructuring agenda, so that the Mexican State could amend it. In addition, Mexico, by its proximity to the United States and Canada, constituted a potential market for the goods produced in those countries or elsewhere with U.S. or Canadian capital. Thus, "[t]he U.S. government. . . began actively to promote the tearing down of barriers to international trade in order to facilitate the globalization of the economy."[78]

The implementation of economic restructuring programs, which included the opening of markets and the liberalization of trade, would potentially benefit developed countries in two ways: first, because Mexico would become a space of production, and second, because it would become a space of consumption. In other words, Mexico represented both a territory whose lands could house different economic activities and an escape valve for those products that could not be sold in the United States and/or Canada. The passage of NAFTA, then, set up the bases not only for the consolidation of the opening of markets in Mexico, but also for the incorporation of the *ejido* and communal lands in the global land market. And when the free trade discourse mentioned Mexico as part of the global economy, it meant that Mexico's land, its resources, and its geography were now entrenched in U.S. and Canadian economies.

THE AGE OF NAFTA

The North American Free Trade Agreement approved in January 1994 by the United States, Canada, and Mexico was the culmination of a series of economic restructuring programs implemented in Mexico.[79] The objectives of this agreement were to reduce tariffs for foreign-produced items, promote export-production for international trade, and smooth the free movement of capitals. Free trade required "a world in which capital is highly mobile and products are exchanged at every step of the production process."[80] In return, NAFTA promised "gains from free trade for all participants" ranging from an efficient and profitable economy to a positive transfer of technology.[81]

The propaganda that sustained NAFTA affirmed that this agreement would accelerate the "integration" of the Mexican economy with that of the United States and Canada, and consequently, (and magically, too) would make Mexico and Mexicans part of the First World (*sic*). The arguments never mentioned the quality of the integration of the Mexican economy nor that of Mexicans into the First World. The statement, however, implied a consideration of the "inferiority" of the development model followed by Mexico and all Latin America, versus that of the United States and Canada. Escobar summarizes that view from rich countries in trying to "develop" and "homogenize" other countries as:

> The intent was quite ambitious: to bring about the conditions necessary to replicating the world over the features that characterized the 'advanced' societies of the time—high levels of industrialization and urbanization, technicalization of agriculture, rapid growth of material production and living standards, and the widespread adoption of modern education and cultural values.[82]

Previous to the signing of NAFTA, critics of structural adjustment programs had advanced the concern that the historical contexts of the participant countries and their capacity to deal with the effect of such restructuring were ignored in designing those policies. And the ignorance of this historical context seemed to be blatant as the application of the structural adjustment programs did not take into consideration regional, ethnic, and cultural differences in Mexico.

Furthermore, as previously stated, so far no evidence existed that those policies, including free trade, worked at all.[83] Opponents to NAFTA emphasized that the treaty was dominated by U.S. and Canadian interests[84] and predicted negative effects from downsizing of firms and corporations, widespread unemployment in the three NAFTA countries, relocation of firms to Mexico motivated by the cheap land and labor, lax environmental regulations, and a radical restructuring of the economic base in the three countries.

In Mexico, the main arguments hoisted against NAFTA were that as agricultural production, policies and land tenure was transformed, rural populations would be once more expelled from the countryside, sending people to urban areas in both Mexico and the United States, and increasing the urbanization of the periphery of cities. Even strong supporters of NAFTA had agreed that there was "an ethical problem arising out of the acceptance of sustained poverty caused by growth and structural change, and this problem was far more acute in Mexico than in Canada and the USA."[85] In sharp contrast, supporters of free trade considered that "[i]nflows of foreign capital continue to be required to fully exploit the growth potential of major Latin American debtor countries."[86]

Again and again, the supporters of NAFTA proclaimed that this agreement would further integrate the Mexican economy, not only to the United States and Canada, but to the "global" economy. This economic integration would benefit populations in both urban and rural areas as the invisible hand of the market pushed them to find their comparative advantage and, consequently, their most profitable economic activity. Thus, as Otero stated ". . .the agreement would consolidate the "neoliberal ideology and the forces of global capitalism in Mexico."[87]

GLOBALIZATION OR IMPERIALISM, OR GLOBALIZING IMPERIALISM?

The theory underlying the logic of international free trade was based on the concept of "comparative advantage "a term coined by Ricardo in 1817.[88] This concept stated that "international trade was based on an absolute advantage, that is, on an exporter with a given amount of resources being able to produce a greater output at less cost than any competitor."[89] The signing countries of NAFTA assumed that each one had a comparative advantage from which to profit as members of NAFTA.

Historically, comparative advantage and free trade have led to an international division of labor[90] that has assigned uneven economic roles to the participating countries and changed their social and cultural structures. Thus, comparative advantage and free trade would industrialize some countries, while others would produce raw materials.[91] In other words, some countries would produce high-value commodities, while others would produce cheap commodities. In consequence, their income and their levels of development, measured within that paradigm, would confine some countries to continue being "developed," while some others would remain "underdeveloped."

Yet, economists tend to assume that comparative advantage forces the design and implementation of policies leading to the organization of international production.[92] In reality, both the organization of production and the policies that sustain it are fostered by transnational corporations and by international and national financial institutions. By this means, nation States have been reduced to a conglomerate of institutions that serve to facilitate the process of international trade through domestic policies and international agreements.

International financial institutions have become the dictators of macro economic policy and planning for Mexico and the rest of Latin America. The role of those institutions has trespassed national and transnational limits causing them to be referred to as "supranational institutions."[93] In reality, the transgression of such limits has, in the vast, been called imperialism.[94] Words matter. Supranational institutions and globalization are new words that have been substituted for imperialism and intervention.

The word "globalization" though sounding benign and even exalting, obscures the tactics through which the process has been implemented in

Mexico. In contrast, the word "imperialism" clearly describes a power relationship between colonized and colonizer. Globalization and imperialism seem like different processes until one compares the features of both and discovers their similarities.

According to Lenin, imperialism "embrace[s] the following five essential features." And he lists:

1. The concentration of production and capital developed to such a high stage that it created monopolies which play a decisive role in economic life.
2. The merging of bank capital with industrial capital, and the creation, on the basis of this "finance capital," of a "financial oligarchy."
3. The export of capital, which has become extremely important, as distinguished from the export of commodities.
4. The formation of international capitalist monopolies which share the world among themselves.
5. The territorial division of the whole world among the greatest capitalist powers is completed.[95]

Since globalization is the buzz world of the moment, a multitude of literature has been written to explain its different aspects and manifestations. For those making policy in the "developed" countries, globalization is seen exclusively as the process by which the planet is becoming a single world through the process of global production and global communications, which results in global citizens.[96]

A different description attempts to merely list the general economic, political, and social transformations currently taking place in the world. According to this description, the characteristics of "globalization" are:

1. Extensive mergers of transnational corporations and alliances of national and transnational companies all over the world.
2. The directing role of international financial institutions in affecting political and institutional changes as well as economic restructuring in developing countries by controlling the lending process of financial capital.
3. The role of free trade agreements in eliminating barriers to private investment which allows the unrestricted movement of transnational capital.
4. The privatization ideology that is attempting to homogenize the cultural economic idea of societies as well as that of their individuals.
5. The formation of regional trading blocs which share among themselves the regional markets under their jurisdiction.
6. Accelerated urbanization that is dividing and reducing rural and Indigenous spaces in both developed and developing countries, and proletarianizing the labor force.
7. Widespread international migration and the creation of Third World enclaves within First World spaces.[97]

TABLE 4.1   Characteristics of Imperialism and Globalization

Imperialism	*Globalization*
Concentration of production and capital (monopolies)	Mergers of transnational and national corporations (monopolies, too)
Finance capital	Increasing influence of transnational capital and supranational financial institutions
Export of capital	Free trade to allow the free movement of capital
Distribution of world markets	Free trade to form regional economic blocks that distribute their markets
Division of world territory	Investment goes to certain regions, creating a "new" geography

Source: Elaborated by author based on Lenin's Imperialism and own notes.

Table 4.1 summarizes and compares the main characteristics of the imperialism defined by Lenin and the globalization process of the present with the purpose of showing their similarities.

To this effect, I conclude that globalization is another wave of imperialism and that we can observe it in the restructuring of the Mexican economy and specifically in the transformation of Article 27 and the *ejido* system. My analysis builds on the works by Addo, Braudel, and Jacobs.[98]

Addo maintains that imperialism is a permanent stage of capitalism and he asks: ". . .why should we choose a new term, when one already exists that covers the essentials of what we are dealing with?"[99] Meanwhile, Braudel states when referring to the characteristics of capitalism:

> Naturally, it is obvious that capitalism today has changed its size and proportions fantastically. It has expanded in order to remain on the same scale as basic exchanges and financial resources, which have likewise grown fantastically. But, *mutatis mutandis*, I do not think that there has been a complete change in the nature of capitalism from top to bottom.[100]

In order to support his thesis on the nature of capitalism, Braudel offers "three pieces of evidence":

1. Capitalism is still based upon exploiting international resources and opportunities; in other words, it exists on a world-wide scale, or at least it reaches out toward the entire world. Its current major concern is to reconstitute this universalism.
2. Capitalism still obstinately relies upon legal or de facto monopolies, despite the anathemas heaped upon it on this score. As they say today, 'organization' keeps circumventing the market. But it is erroneous to believe that this is anything really new.
3. Furthermore, despite what is usually said, capitalism does not overlay the entire economy and all of working society: it never encompasses both of them within one perfect system all its own. The triptych I have

described—material life, the market economy, and the capitalist economy—is still an amazingly valid explanation, even though capitalism today has expanded in scope.[101]

Both Addo's and Braudel's notions of imperialism and capitalism complement each other. Addo emphasizes that: "The connection between imperialism and world capitalism is so much an intimate part of world capitalism that capitalism cannot exist without imperialism. We may in fact see the two phenomena as synonymous."[102] While Braudel states: ". . .capitalism has always been monopolistic, and merchandise and capital have always circulated simultaneously, for capital and credit have always been the surest ways of capturing and controlling foreign markets. Long before the twentieth century the exportation of capital was a fact of daily life."[103]

Therefore, when analyzing the nature of the intervention of supranational entities like the World Bank, the IMF, and private banks in dictating the direction of Latin American policies it might be pertinent to use the term that "already exists [and] that covers the essentials of what we are dealing with"[104]. Fanelly, et al report on the nature of that intervention and describe how the meaning of instability was determined by "Washington" and how the Fund and the Bank penalized "debtor countries" that did not follow "stabilization policies."[105] In that context, instability was understood as the deficit created by profuse borrowing to finance the public sector. Debtor countries structuring in order to straighten out their economies were eligible to obtain new credits; however, access to credit was blocked when their policies were unsuccessful or were not adopted.

The fact that structural adjustment programs are instigating a new wave of privatization of Indigenous lands and displacement of their communities call attention to Jacobs's work that affirms that "[t]he economic dimension of imperial expansions (be they colonial or 'newly' global) is undeniable, as are the uneven divisions of power and privilege they produce."[106]

Those uneven divisions of power and privilege were produced in Mexico between the colonizer and the colonized, the Indigenous populations and the *Criollos* and well-off *Mestizos*. The history of land in Mexico is a history of privatization of Indigenous lands which can be defined in the words of Said as encompassing a continuous "struggle over their territories."[107] That is why the Zapatistas in Chiapas declared war, because they were struggling to preserve their territories, their lands.

To describe the current imperialism, Jacobs insists that "[t]his is not simply a politics that is against globalization. Nor is it simply a return to origins. Imperialism, in whatever form, is a global process—it occurs across regions and nations—but even in its most marauding forms it necessarily takes hold in and through the local. The embeddedness of imperialist ideologies and practices is not simply an issue of society or culture but also, fundamentally, of place."[108] The next three chapters expose struggles over territory at the local level of the *ejido* as place.

Chapter Five

San Luis Río Colorado:
The First *Ejido* to Privatize its Land

San Luis Río Colorado was the first *ejido* that officially privatized land under the 1992 reforms to Article 27.[1] It was also the first one to constitute a so-called *ejido* real estate partnership with foreign private capital. The official bulletins that the *Procuraduría Agraria* published depicted it as the successful example of what *ejidatarios* could achieve by privatizing their lands. Although San Luis Río Colorado (SLRC) was portrayed as a model of success, it was not really fulfilling the pretended modernization of the agri-culture sector in Mexico, of which supporters of the Reforms to Article 27 had boasted during the approval of these amendments. The investment that this *ejido* attracted was converting its agricultural lands into a mega industrial park with real estate development for the future establishment of additional *maquiladoras*,[2] industry that mushroomed in the Northern region of Mexico due to its strategic proximity to the United States.

This chapter presents the case of the San Luis Río Colorado *ejido* during the time of its privatization. It shows the *ejido* from within and explores its relationships with the city and the region where the *ejido* is located.

LOCATION AND DESCRIPTION

San Luis Río Colorado (SLRC) is located in the border state of Sonora in Northwestern Mexico with an area of approximately 14 hectares. This is a prime location because the *ejido* is adjacent to the state of Baja California in Mexico, and to California and Arizona in the United States. In this sense, SLRC is part of the Northwestern border region that comprises the Mexican cities of Tijuana and Mexicali and the U.S. cities of Yuma, Arizona, and San Diego, California (Map 5.1).

The population of the San Luis Río Colorado *ejido* is composed mostly of several waves of immigrants from other Northwestern states. The number of Indigenous people who inhabited the region has decreased considerably. From the more than eleven Indigenous groups living in SLRC, the 1990

Map 5.1   Sonora, San Luis Río Colorado, and the Northwestern Border. *Source:*
Drawn by the author.

census reported that just 0.4 percent spoke an Indigenous language.
Because of its proximity to the border, immigrants gradually populated
that area after the 1910 revolution. Later, the *Bracero* Program[3] and the
Border Industrialization Program, commonly known as the *maquiladora*
program,[4] contributed to further in-migration to the area.

The *Bracero* Program started in 1951 and allowed U.S. farmers to temporarily employ Mexican labor.[5] In 1964, when Mexican labor was not
needed, the United States ended the program. Some of the *braceros*
remained in the United States, others returned to Mexico, and some others
remained on the Mexican side of the border.

The *maquiladora* program began in 1965. Its focus was to assemble
parts for export. The *maquiladora* program absorbed some of the labor
force that the end of the *Bracero* program had left behind, but mostly
attracted internal migration from nearby states. It is interesting to note
that *maquiladoras* were located in "restricted areas"[6] which, according to

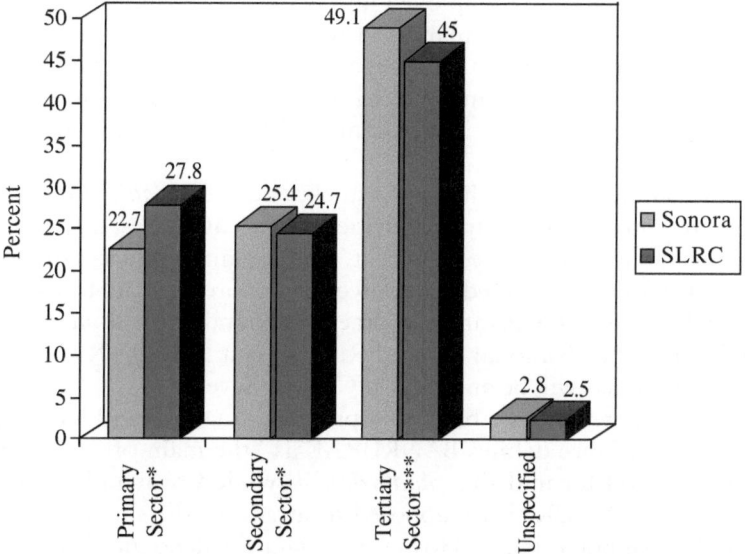

Figure 5.1 Employed Population by Economic Sector in Sonora and San Luis Río Colorado, 1990. *Includes agriculture, livestock, forestry, and fishing. **Includes mining, oil and gas extraction, manufacturing industry, generation of electric energy, and construction. ***Includes commerce and services. *Source:* Instituto Nacional de Estadística, Geografía e Informática (INEGI), *San Luis Río Colorado, Estado de Sonora. Cuaderno Estadístico Municipal. Edición 1994* (Aguascalientes, México: INEGI, 1995), 56. Based on data from INEGI, "Sonora, Resultados Definitivos. XI Censo General de Población y Vivienda, 1990."

Article 27, did not allow the presence of foreign capital. The Mexican government appears to have made sure to "exempt [*maquiladoras*] from the Mexican laws requiring majority Mexican ownership."[7] Most of the *maquiladora* industry has located in the Northern border region in cities such as "Tijuana, Mexicali, Nogales, Ciudad Juárez, and Matamoros."[8] The San Luis Río Colorado municipality encompasses 41 *ejidos*, the SLRC *ejido* being only one of them.[9] Currently, approximately 24 *maquiladoras* have been established in this municipality. Over time, with migration and the pull of the *maquiladora* industry, the SLRC *ejido* has become a city of approximately 135,000 people.[10]

LOCAL ECONOMY OF SAN LUIS

According to the 1990 Census, both at the state and the municipality levels, almost half of the population is employed in the tertiary sector. As indicated by Figure 5.1, 49.1 percent of the state working population and 45 percent of the municipal labor are employed in the tertiary sector. The statistics in this figure show that the primary and secondary levels employed almost the same populations both at the state and municipal levels, with a slightly higher percentage in the primary sector of San Luis Río Colorado, which implies that the SLRC municipality continues having a strong agricultural sector.

Export-commercial agriculture, livestock, and the *maquiladora* industry are the predominant economic activities in the area. Sesame, soybeans, cotton, vegetables, and, above all, wheat, are the main products grown in the *ejido*. SLRC had access to credit, resources, and more capacity to work independently from the bureaucratic and praetorian hand of the *Banco Nacional de Crédito Rural* or National Bank of Rural Credit (BANRURAL) than the Ixtaltepec *ejido* that will be analyzed in Chapter Seven.

BANRURAL was a State bank that provided financial and loan support to rural areas. Myhre defines BANRURAL as "the main official source of agricultural credit for mid-size *ejidatarios*,"[11] while Covarrubias Patiño indicates that "BANRURAL is supposed to attend to the small landholders with productive potential."[12] However, *ejidatarios* denounced the timeless awarding of loans and credits, the "extensive paperwork"[13] to apply for financial assistance, and the obscure practices of both *ejido* and BANRURAL officials.[14] Thus, when BANRURAL attempted to force *ejidatarios* from SLRC to buy their agricultural inputs from a certain store in exchange for approving loans for them, *ejidatarios* disassociated from the Bank. This separation shows the degree of independence, resources, and decision-making of SLRC which was nonexistent in other *ejidos*.

Although the *Procuraduría Agraria* reported in 1994 that 90 percent of economic activity in the *ejido* was related to agriculture or livestock, census data for the SLRC municipality indicated that only 34.1 percent of the *ejido* was devoted to agriculture, 4.9 percent was allocated to livestock, and a growing 61 percent was classified as partaking of other activities (see Figure 5.2).

These contradictory data were clarified when the Procuraduría Agraria document specified that "from all that land, only 1,673 hectares are cultivated . . . "[15] meaning between 11 and 13 percent of total land in SLRC, a figure that does not make sense if 90 percent of the population were farmers. The high percentage corresponding to "other activities" in the municipality was related to the accelerated growth of the manufacturing sector in the region. In 1992, the *maquiladora* sector had reported that, in SLRC alone, there were 22 *maquiladora* industries. Thus, it is likely that those industries account for a good proportion of the 61 percent of the total economic activity in the *ejido* classified as "other."[16] In comparison to the state data, it was evident that SLRC had undergone a transformation. At the state level, primary activities in *ejidos* accounted for 86.1 percent—disaggregated as 47.5 percent in agriculture, 36.1 percent in livestock, 1.5 percent forestry, and 1 percent gathering (see Figure 5.2).

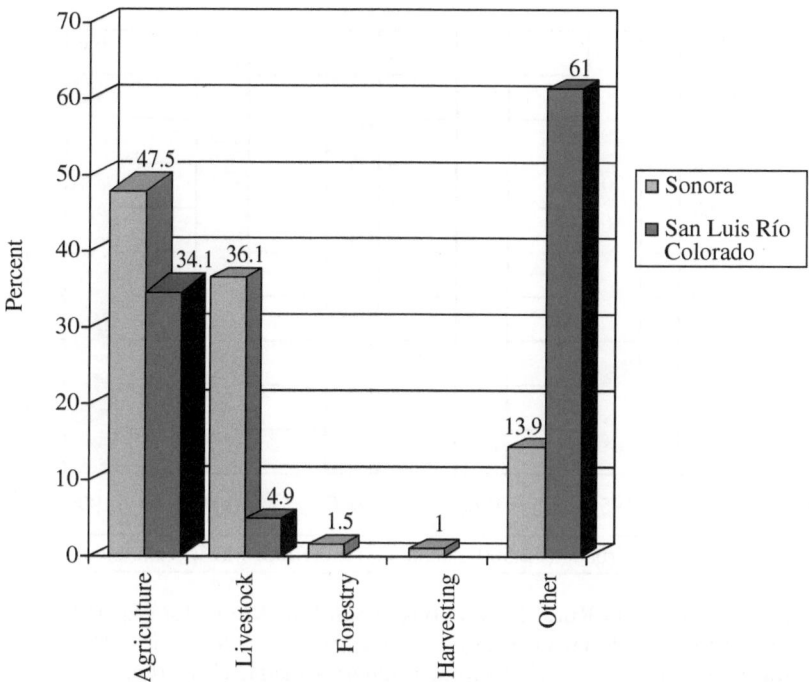

Figure 5.2   *Ejidos* and Agrarian Communities by Main Economic Activity in
Sonora and San Luis Río Colorado, 1991. *Source:* Instituto Nacional de Estadística,
Geografía e Informática (INEGI), *San Luis Río Colorado, Estado de Sonora.*
Cuaderno Estadístico Municipal. Edición 1994 (Aguascalientes, México: INEGI,
1995), 65. Based on data from INEGI, "Resultados Definitivos. VII Censo Ejidal,
1991."

Meanwhile, in SLRC primary activities accounted for 39 percent. In
contrast, at the state level, the share of other activities in *ejidos* was only
13.9 percent while in SLRC it was 61 percent. This could be an indication
of the transformation of *ejido* land to other uses, which seems to be corroborated by the data on rural and urban populations in the municipality
(see Figure 5.3).

According to this figure, in 1950 the population of SLRC was 70 percent
rural and 30 percent urban; in 1990 the rural population was only 9.9 percent
and the urban population was 90.1 percent. This data should be read with
reservations as the definitions of rural and urban are based on localities of
less and more than 2,500 inhabitants, respectively.

This definition does not take into consideration the fact that the number
of inhabitants in the Northern region does not necessarily determine the
urbaness or ruralness of an *ejido*. Big extensions of *ejido* land can be considered either rural or urban, in spite of the number of their inhabitants.

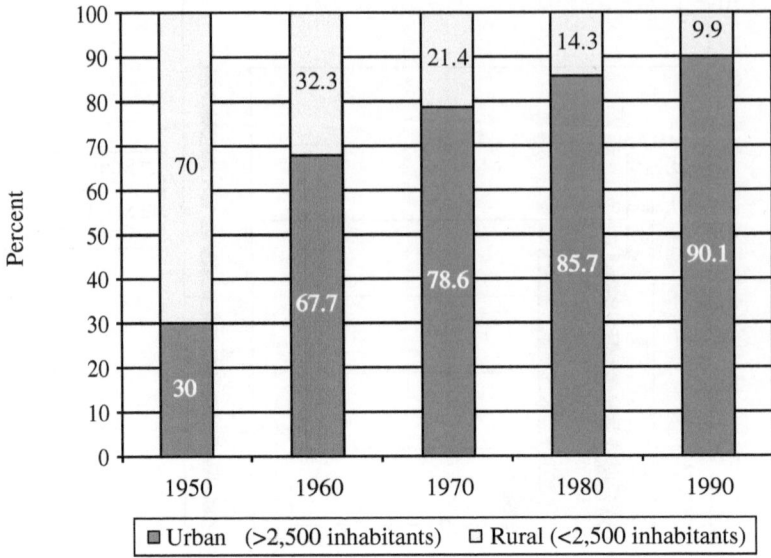

Figure 5.3 Urban and Rural Populations in San Luis Río Colorado, 1950–1990. *Source:* Instituto Nacional de Estadística, Geografía e Informática (INEGI), *San Luis Río Colorado, Estado de Sonora. Cuaderno Estadístico Municipal. Edición 1994* (Aguascalientes, México: INEGI, 1995), 18. Based on data from INEGI, "Sonora, Resultados Definitivos. VII–XI Censos Generales de Población y Vivienda, 1950–1990."

Considerations of economic activity, lifestyle, culture, and identity are not part of the definition of what rural or urban are, but they should be. The concepts of urban and rural must change as land and their local economies change.

Wage statistics in the municipality of SLRC (see Figure 5.4) indicate for 1990 that 9.3 percent of the working population makes less or no minimum wage (8.1 percent receive less than one minimum wage and 1.2 percent receive no wages), while 33 percent of the population receives between one and two minimum wages, which means that 43.3 percent of the population are low income. This figure also shows the marked regional differences in wages as, compared to other regions, SLRC has low unemployment, and 56.7 percent of its population receives 3 times the minimum wage or more.[17]

HISTORY OF THE *EJIDO*

SLRC was established as an *ejido* in February of 1929. Originally 86 *ejidatarios* were given 1,731.69 hectares. In the 1930s, during the administration of Lázaro Cárdenas, 14,900 hectares more were granted to 148 *ejidatarios*. A 1994 official document by PROCEDE specified that the *ejido* had 236 *ejidatarios*, but did not specify the exact amount of total

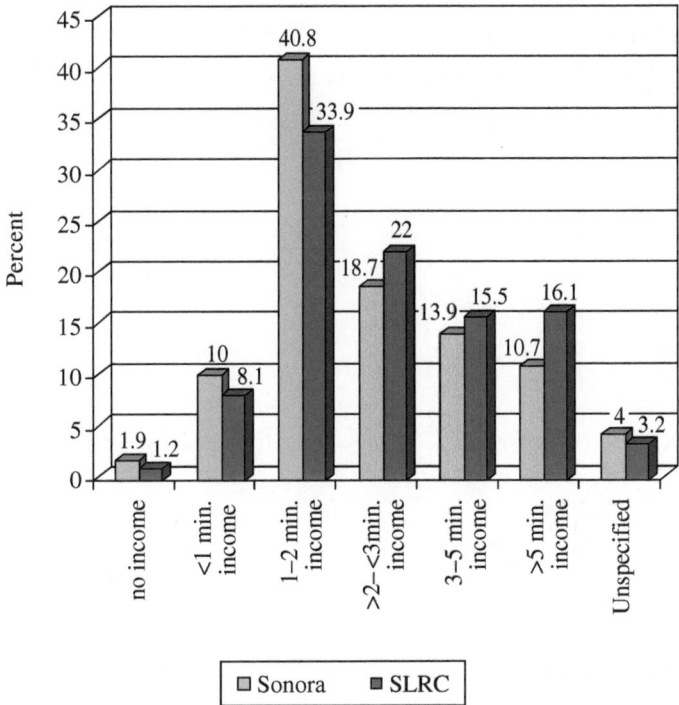

Figure 5.4 Employed Population by Monthly Income in Sonora and San Luis Río Colorado, 1990. *Source:* Instituto Nacional de Estadística, Geografía e Informática (INEGI), *San Luis Río Colorado, Estado de Sonora. Cuaderno Estadístico Municipal. Edición 1994* (Aguascalientes, México: INEGI, 1995), 57. Based on data from INEGI, "Sonora, Resultados Definitivos. XI Censo General de Población y Vivienda, 1990."

ejidal land. Instead, it gave the extension of "between 13,000 to 15,000 hectares."[18] And it explained, "[t]he exact figure, whatsoever, has never had much importance since, for the most part they [the lands] are barren, desert-like and unproductive for agriculture and even for livestock."[19] These "barren" lands, however, bordered the states of California and Arizona, a proximity that made them valuable in spite of not being appropriate for agriculture.

During my first visit to this *ejido*, the deep differences existing among *ejidos* and regions within Mexico became evident. SLRC is not the traditional agriculture-oriented, rural area that other *ejidos* were. Rather, SLRC is a fast-growing border city with approximately 25,000 dwellings and 135,000 people.[20] The city of SLRC had followed the same pattern of urbanization as had Mexico City: it had gradually grown over time to approximately 3,700 hectares of the SLRC *ejido*. Currently, SLRC is an *ejido*/city. Since the lands of SLRC continue being converted from agricultural to *maquiladora*

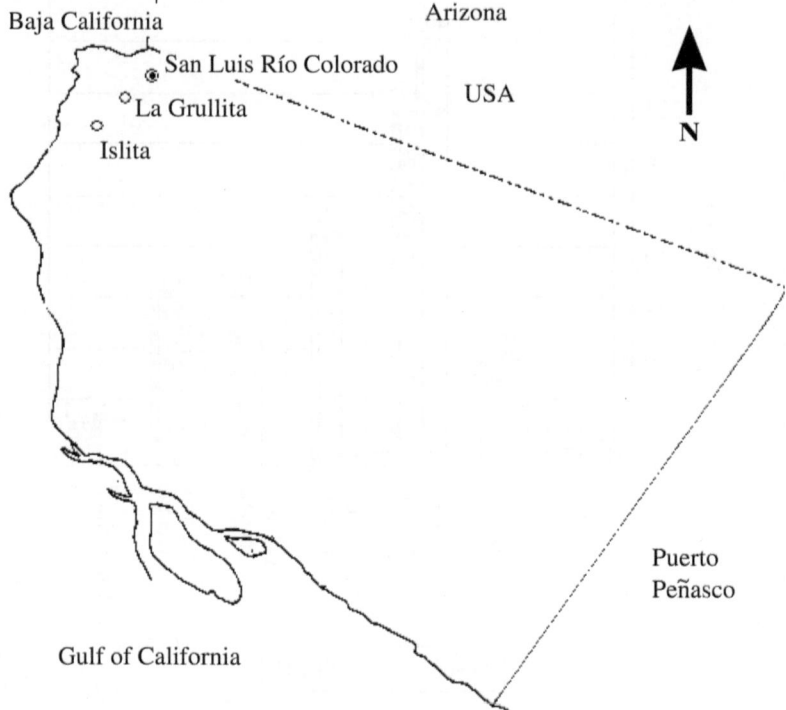

Map 5.2   *Ejido* San Luis and Surrounding *Ejidos. Source:* Instituto Nacional de Estadística, Geografía e Informática (INEGI), *San Luis Río Colorado, Estado de Sonora. Cuaderno Estadístico Municipal. Edición 1994* (Aguascalientes, México: INEGI, 1995), 8. Based on data of the government of the state of Sonora, cartographic map of the state of Sonora.

activities, which have become the dominant regional economy in this border area, we could say that SLRC is an *ejido*/city/region. The *ejido* SLRC has slowly been converting the uses of its land through several privatizations that have changed the space of agricultural activities into space for urban housing and *maquiladoras*.

THE CONSTANT PRIVATIZATIONS OF *EJIDO* LAND IN SLRC

Agricultural land has been decreasing in SLRC because several expropriations have gradually taken place in order to convert *ejido* land into urban private land. These expropriations had the purpose of "regularizing" *ejido* land previously sold for housing and commercial purposes by *ejidatarios*.[21] Sales of *ejido* land quickly urbanized SLRC, transforming the *ejido* into a city. This process of rapid urbanization was reaching other nearby *ejidos*, such as La Grullita and Islita, within the municipality of San Luis Río Colorado (see Map 5.2).

The first post-revolutionary wave of in-migration to SLRC took place during the U.S. economic crisis in 1929. The crisis spurred widespread unemployment of Mexicans in the U.S. and their massive return to Mexico. SLRC was one of the places where the returning populations relocated. The sudden increase in population in the *ejido* put pressure on land and resources. Thus, the *ejido* requested additional land for agriculture first in 1937 and again in 1940 and on both occasions, the requested land was granted.[22] However, it was obvious that during the period 1929–1937, *ejido* land was sold, at least for housing purposes, to the new residents in the area. Evidence of that growth was the creation of the municipality of San Luis Río Colorado in 1939.

Three years later, in 1942, the municipality would be requesting 101 hectares for the creation of its *fundo legal*, or legal land patrimony, which is the land legally assigned to establish the town of the *ejido*. As population growth accelerated, new expropriations of *ejido* lands were approved. In 1950, 630 hectares were expropriated and in 1963, 297 more hectares became part of the *fundo legal*. By 1992, the legal extension of the *fundo legal* was 1,028 hectares. This area did not comprise the total urbanized area that, in very conservative terms, the PROCEDE bulletin on SLRC accounted as 3,700 hectares.[23] This number is an estimate of the total *ejido* land illegally sold for non-agricultural purposes before the *Ejido* Reform in 1992, but it does not include the area that did not undergo regularization. Thus, *ejido* land privatization and its use conversion were not new processes in SLRC, since they have been occurring constantly since 1929.

THE FIRST PRIVATIZATION OF *EJIDO* LAND UNDER THE 1992 ARTICLE 27

"It can be said that we reinvented Article 27." These were the words of Enrique Orozco Oceguera, representative of the mercantile society created with the privatized *ejido* land, during an interview about the industrial park project that foreign capital and *ejidatarios* in SLRC were launching. During our interview Mr. Orozco Oceguera mentioned that two years prior to the joint venture, "they had everything prepared."[24] According to Orozco Oceguera, the project started when a visiting Canadian entrepreneur was looking for 100 hectares of land in Mexico. Orozco Oceguera volunteered to conduct a search in his native SLRC. When the visiting entrepreneur realized that SLRC already had an industrial park as well as U.S. firms located in the area, he and his colleagues enlarged their project to develop an industrial park of at a least 5,000 hectares.

The industrial park that had impressed the foreign capitalists was the *Promotora Industrial de SLRC* (Industrial Promoter of SLRC), established in 1980 as a public-private industrial park. In 1986, Gustavo Garza wrote: "The one [park] in San Luis Río Colorado has been only a year in operation and it will not be easy to achieve a continuing installation of four firms annually."[25] Time proved him wrong as by 1994 there were at least 22 *maquiladoras* operating in the area. This already-operating industrial development encouraged the location of the new international industrial park in 1994 which would, in turn, give birth to the *Constructora e Inmobiliaria Ejido San Luis* (COINE) (*Ejido* San Luis's Building Contractors and Real Estate Development). The use of the word *"ejido"* in the title of the industrial park project hides the fact that once *ejido* lands were privatized, they stopped being legally considered as *ejidos*. Instead, they had become private property.

Orozco Oceguera told me that before Article 27 was modified in 1992, in order to find the legal mechanisms to convert the *ejido* land necessary for the new industrial park into private land for the location of the international industrial park, he had been negotiating an expropriation of *ejido* land for industrial purposes with several ministries in Mexico City. When foreign capitalists and Orozco Oceguera learned about the 1992 *Ejido* Reforms they decided to wait until they had them approved. In this way, they avoided the expropriation mechanism and initiated the first legalized privatization of *ejido* land protected by the 1992 Article 27.

Had Orozco Oceguera followed the expropriation path of negotiation, land would have been in the hands of *ejidatarios* in the form of an *ejido* enterprise, instead of that of a mercantile society as permitted by the 1992 Article 27. The difference between the *ejido* enterprise and the mercantile society lies in tenure. An *ejido* enterprise belongs to the *ejido* community and is regulated by Agrarian Law; a mercantile society is exclusively private property and is regulated by private law legislation, which treats its members as stockholders. To this effect, when asked about who were the new decision-makers in the new society, Orozco responded: "It is not appropriate to ask who decides. There are many questions that are inappropriate to ask." Then he added that the Nafta group, that is, the joint venture mercantile society, was the one who decided.[26]

NAFTA AND COINE

The term "Nafta group" has nothing to do with the acronym of the North American Free Trade Agreement (NAFTA). Orozco chose the name Nafta as the propagandistic title of the new mercantile society. The Nafta group encompassed on the one hand the investor countries: Canada, Japan, Spain, and the United States, and, on the other, COINE, which represented the ex-*ejido* owners—now stockholders—under their new status as private proprietors.

Originally, *ejidatarios* were planning to form an *"ejido* enterprise" in conjunction with foreign capital. However, in order to form such a joint enterprise, foreign capitalists required that *ejidatarios* privatize their land. For that reason *Constructora e Inmobiliaria Ejido San Luis* (COINE) was created.[27] COINE would become the association through which *ejidatarios* would hold privatized *ejido* land. For purposes of the enterprise, COINE was requested to transfer the privatized *ejido* land to the Nafta group for a period of 99 years, the span of time that the Nafta group expected to last.

The main objective of the Nafta group was to acquire *ejido* land in SLRC in order to build an international industrial park. Basically, the Nafta group was the mercantile society that was going to plan, develop, and administer the lands. It was also going to build or subcontract to build infrastructure for the industry locating in the park. In addition, they could sell or rent land. This latter function was very ambiguous as *ejidatarios* could just as well have directly sold or rented their land without having to transfer it to a mercantile society that would then function as an intermediary between the ex-*ejidatario* and the capital requesting land for industrial uses.

Curiously, the Nafta group was described by the Procuraduría Agraria as the "*Ejido* Enterprise";[28] something it was not. It had in fact become a mercantile society. The Nafta group was a private society holding privatized *ejido* land which, since its privatization, was no longer *ejido* land. Neither the land nor the *ejidatarios* were protected any longer by *ejido* legislation, because, by forming the "*ejido* enterprise" which was in reality a mercantile society, *ejidatarios* had given up their status as such as well as their direct ownership of *ejido* land, in favor of the Nafta group. The official discourse of the Procuraduría Agraria disguised this situation and created conflict in the *ejido* as *ejidatarios* kept on calling themselves *ejidatarios* and wanted to keep on being treated as such, although in fact they were private proprietors.

In addition to transferring the land to the Nafta group, each "*ejidatario*" was required to buy a \$100N[29] share in the enterprise as a requirement to become shareholders. For some reason that nobody could explain, thirteen "*ejidatarios*" were awarded the preference of buying five shares in addition to the one they had acquired. The material from the *Procuraduría Agraria* does not specify who those *ejidatarios* were and the reasons why they were allowed to acquire more shares than the others. In any case, the *ejido* provided the land for the industrial park and the initial capital for the enterprise. Some *ejidatarios* saw that these requirements added more conflict to the privatization process in SLRC as they considered that providing the land was more than enough.

A DIFFICULT AND DUBIOUS PRIVATIZATION

Privatization of *ejido* land in SLRC prompted confrontations between those *ejidatarios* supporting COINE and those opposing it. This division between *ejidatarios* occurred when the process of privatization of *ejido* land did not follow the regular procedures established by PROCEDE. Apparently, investors and some of the interested *ejidatarios* held meetings without the required quorum, consequently, the measuring and setting of limits were not approved by all the *ejidatarios*. Finally, many arrangements were not clear between COINE and *ejidatarios,* nor between COINE and the Nafta group.

For example, in terms of participation, *ejidatarios* held 25 percent of the decision-making power and representation, while foreign investors held 75 percent. The Nafta group did not comply with foreign investment regulations nor with the mercantile societies' legislation, since the participation of foreign investment surpassed the 49 percent stipulated by law. In addition, each *ejidatario* shareholder did not count as a full vote, as the law read.[30] Instead all *ejidatarios* represented just 25 percent of the voting power. Evidently, in a voting situation, *ejidatarios* would always be a minority, even though they had contributed the initial capital and all the land for the international industrial park. In spite of such an obvious bias, the Nafta group had been registered as a Mexican society in order to circumvent the law. But the irregularities were many. A final one was the clause that stated that should the society between *ejidatarios* and private foreign investors be dissolved, the *ejidatarios* would receive 25 percent of the capital, as long as all the debts of the society had been paid.[31] Thus, *ejidatarios* were carrying all the risk of the enterprise.

Ejidatarios' discontent emerged, in part, as a result of the unfair distribution of costs and benefits. According to the contract with foreign investors, they had to pay for the total amount of taxes and government fees, while they would get only 25 percent of the benefits of the *"ejido* enterprise.*"* *Ejidatarios* scoffed at the *ejido* enterprise as they had been told that, by organizing such an enterprise, they would become entrepreneurs of their own land. However, under the new arrangement that converted them into private proprietors, they did not own their land since it had been transferred to the Nafta group. Nor could they make any claims to their land because, by participating in the *ejido* enterprise, they had lost all their rights as *ejidatarios*. They were not at all familiar with the status of private proprietor, but they were well aware they did not like it. An *ejidatario* told me they were very angry because "They do not treat us as *ejidatarios* anymore."[32] He meant that they were not considered as owners of their land anymore.

Deep conflicts of interest were evident among *ejidal* authorities and the representatives of foreign capital. The most visible case pointed to the person who was the president of the *Comisariado Ejidal*. During the privatization process, Rafael Meza, in his role as an *ejido* authority, had to sign all the corresponding documents to approve, in name of the *ejido*, the privatization of land. In the process of doing so, he transferred the ownership of land to himself as the newly appointed president of COINE. Since COINE had to transfer the privatized land to the Nafta group, he also signed the corresponding documents. Interestingly, on the side of the Nafta group, he was also the person receiving the transferred land as he was the vice president of Nafta. Suspicious *ejidatarios* mentioned that Meza had once held an official public position in CORETT, which was the government agency in charge of expropriating land to regularize it.[33] They noted that he was well versed in all these processes of land transfer and privatization and they distrusted the fact that a single person in the *ejido* had acquired so much decision making power in the "*ejidal* enterprise." This apparent conflict of interest might have prompted the angry response by Orozco Oceguera to my question about who decided in the new society: "It is not appropriate to ask who decides."

It seemed to be a very dubious and difficult privatization process. The changes in legislation, the conflicts of interest, the fees to be paid here and there, had made *ejidatarios* doubt that the project of the international industrial park would bring any benefits to them. This situation was aggravated by the fact that the representative of the Nafta group, Enrique Orozco Oceguera, had told *ejidatarios* that any real benefit would not be visible and payable for a period of at least 15 years. In a personal interview, an 80 year old *ejidataria* told me she was not going to live to see the benefits that the Nafta group proclaimed. She was one of the participants in the *ejidal* enterprise.[34]

TRANSNATIONAL COMMUNITIES

During my first week in SLRC, I stayed at Ramona's home. Ramona was the Treasurer of the *ejido* San Luis Río Colorado. A friend of hers allowed her to share this house whenever she was in San Luis Río Colorado. Ramona, like some other *ejidatarios* in the region, had made her permanent home in Yuma, Arizona, a city across the border from San Luis Río Colorado. Ramona is called a *rodina* or *emigrada* because she, like other people in SLRC, holds a U.S. green card. She had secured it during one of the amnesty programs to obtain legal residence in the United States. Her permanent residence is in Yuma, Arizona, but she owned *ejido* land in SLRC. This situation is not uncommon since most people in SLRC have relatives living and working in Arizona or California who properly fit the definition of transnational communities—those who live in one country and work in another one, as well as those who commute back and forth between countries. According to the results of my field research, *emigrados* typically worked as farmworkers in California or in the *maquiladora* sector.

This transnationality, however, poses the problematic question of who should own land, especially if it is *ejido* land. The revolutionary slogan of "redistribution of land for those who work it" crashes here on the border. Many *ejidatarios* in SLRC belong to a very different profile of *ejidatario*. Most of them do not work the land, many were beneficiaries of *ejido* land because somebody else had included them in the list of names of those organizing to request *ejido* land. As *ejido* land urbanized in this area, many *ejidatarios* have used it for speculation. The fact that the border is a transitional area between countries, gives a different character to the question of ownership of land. This area comprises a very different Mexico from that of Ixtaltepec, Oaxaca or La Poza, Guerrero, the *ejidos* that I examine in Chapters Six and Seven.

Since the population of the ex-*ejido* kept on calling themselves *ejidatarios* and the land as *ejido*, it demonstrates that the *ejido* is also the community of people organized around the space that the *ejido* provided, which is one of the multiple representations of this *ejido*, that has now become a border city.

THE *EJIDO*-CITY

In order to apply the structured interview that I had prepared for this research, I tried to go home by home, as I had done in the other two *ejidos* of this research. The task proved impossible as the *ejido* was a city of 135,000 inhabitants. Consequently, most of the people living in the *ejido* were not *ejidatarios* of San Luis Río Colorado. Because of the periodic expropriations of *ejido* land that CORETT[35] had undertaken, several parts of the *ejido* had already been privatized for urban purposes to the point that the expropriated land reached in 1995 at least 45 percent of the total *ejido* land.

Thus, people living in the city of San Luis were not or had not necessarily been *ejidatarios*. Most of them had bought land from *ejidatarios* or purchased it from the government when expropriations had taken place. Some others possessed *ejido* land in other *ejidos* adjacent to SLRC, some rented land in the *ejido* and a very small number of the residents of the *ejido* were "real" *ejidatarios* of SLRC. The *ejidatarios* of SLRC accounted for 0.17 percent of the total population. San Luis Río Colorado was now a city, not an *ejido*. Or, in any case, it was an *ejido*-city as *ejido* land transformation had taken place extensively and land use changes and urbanization had taken over the space of the *ejido*.

Because of the variety of land tenure relations within the *ejido*, the high rate of crime in the urban areas of SLRC, and the extension of the *ejido*, the *ejido* authorities suggested that I start interviewing the *ejidatarios* who came to the *ejido* offices. I did so and every time I had access to transportation I also visited some of the *ejidatarios* at their homes. I found that two of the housing characteristics in SLRC were more resources for building and transnationality. Residents in this part of the country had access to most urban services and housing structures were either of brick and mortar or similar to the wood houses in Yuma. Transnational residents had adopted forms of architecture and materials that are not generally used for housing in Mexico. In addition, the layout of the city had followed the pattern of that of suburban areas in the United States. Horizontality and urban sprawl were characteristics of the more than 24,000 housing units in SLRC.[36]

PROCEDE IN SLRC

The implementation of the PROCEDE program eased my data gathering and interview process as public officials from CORETT had cited all *ejidatarios* of SLRC in the offices of the *ejido*. By the end of 1995, *ejidatarios* were in the last stages of the PROCEDE program and they were ready to receive their *ejido* land certificates and their titles for urban lots. They had also to formalize and ratify the legal successors of their *ejido* land and this latter process had to be carried out in the *ejido* office in person. Thus, I was able to interview them while they were in the office of the *ejido*. The disadvantage of this strategy was that I did not visit every *ejidatario* home of my interviewees and the relationship was very different from the one I established in Ixtaltepec and in La Poza. The advantage was that I could witness the official procedure of awarding of *ejido* land titling and certificating in the first *ejido* that had decided to privatize *ejido* land. In addition, I was also present while boundary and ownership issues emerged even after the process of measuring was over, which questioned the accuracy of the titling program and the unresolved boundary delimitation issues that plagued the application of PROCEDE. Finally, I also observed the way *ejidatarios* were designating their successors or inheritors which had gender undertones as most male *ejidatarios* decided to name as their successors other male relatives, but not their wives or daughters.

THE WELL-OFF *EJIDATARIOS*

The SLRC *ejidatarios* were very different from those in the two other *ejidos* of this research. *Ejidatarios* in SLRC had an office building, with secretaries, conference rooms, phones, and fax machines, among other things. All *ejidatarios* owned a vehicle. These characteristics reminded me anew of a talk with Daniel Coss Rangel, an economist from Hermosillo, who had said: "When one talks about *ejidatarios* in the North, one is referring to well-off landowners who drive a pickup, wear $500 dollar boots, with access to credit, with resources. When one talks about *ejidatarios* in the South, one is referring to poor campesino Indians who do not even have a mule, wear huaraches and have no access to credit."[37] The evidence of that statement was present in the office of the *ejido* SLRC. Indisputably, the *ejidatarios* and their families in Sonora had a radically different profile than those in Ixtaltepec or La Poza.

Even educational levels were higher than those found in the other two *ejidos*. By 1990, Sonora and SLRC had an illiteracy rate of 5.3 percent, one of the lowest in the country. As Figure 5.5 shows, about half of the total population at the state and the municipality levels had high school education or beyond. In comparison, in the *ejido*, 11 percent of the interviewees were illiterate, while 58 percent had had some kind of elementary education and the remaining 31 percent had high school education or beyond.[38] In spite of the fact that the rate of illiteracy in the *ejido* was higher than that of the municipality or the state, it was not as high as I found in the *ejidos* of Ixtaltepec and La Poza.

LABOR AND MIGRATION

Currently, the SLRC *ejido* consists of 270 *ejidatarios*. About 17 percent of them are women. Data obtained by the structured interview undertaken

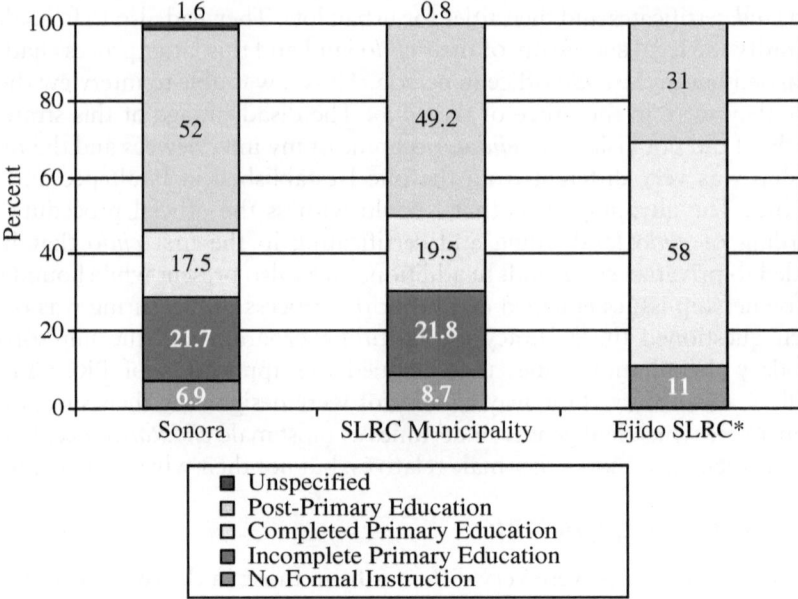

Figure 5.5 Population 15 Years and Older by Educational Level in Sonora, San Luis Río Colorado, and the *Ejido* SLRC, 1990. *Data obtained from the results of the structured interviews carried out in the *ejido* by author in 1994–1995. *Source:* Instituto Nacional de Estadística, Geografía e Informática (INEGI), *San Luis Río Colorado, Estado de Sonora. Cuaderno Estadístico Municipal. Edición 1994* (Aguascalientes, México: INEGI, 1995), 42. Based on data from INEGI, "Sonora, Resultados Definitivos. XI Censo General de Población y Vivienda, 1990."

among the *ejido* population indicated that 71 percent of *ejidatarios* worked in agriculture or in an agriculture-related activity, while only 24 percent of them worked in industry or any other non-agricultural activity. When asked about their previous and current occupations, it was evident that there had not been a significant variation in men's jobs over time, except that 18 percent of those previously working as farmworkers in SLRC, continue doing so either in SLRC, in the surrounding *ejidos* of the SLRC municipality, or in the United States, depending upon the season (see Figure 5.6). This meant that this population became transnational farmworkers with different status and identities. These *ejidatarios* are both owners of *ejido* land and farmworkers in SLRC, while in the United States, they are migrant farmworkers.

In this *ejido* it seemed out of place to ask questions about firewood and drinking water since most of SLRC was urbanized. This urbanization is reflected in the labor markets dominant in this border region. In San Luis

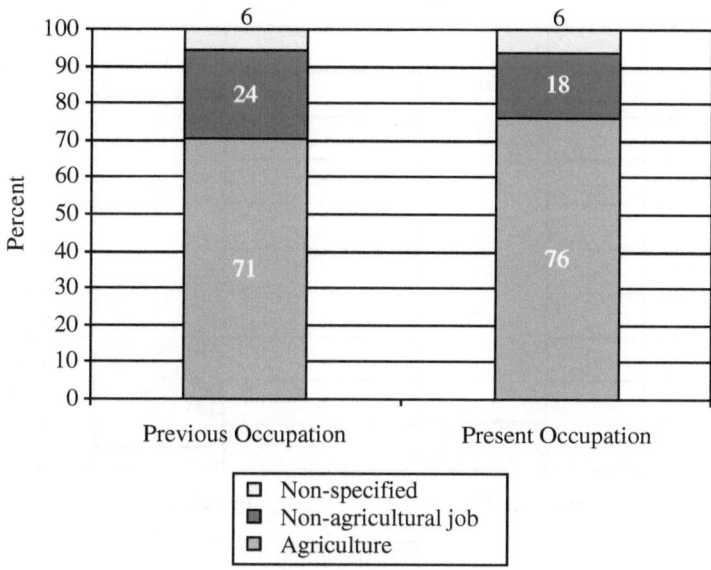

Figure 5.6 Previous and Present Occupations of the Male Labor Force in SLRC, 1994. *Source:* Structured interviews carried out in the *ejido*.

alone, there are about 24 *maquiladoras* that employ a large number of young women. In contrast to men, women in the *ejido* tended to participate more in the labor markets generated by the industrial sector. Thus, a process of feminization of the labor force in industry occurred here.

Figure 5.7 shows that 42 percent of women in the *ejido* worked in non-agricultural jobs such as *maquiladoras* and food preparation, 37 percent were housewives, 16 percent worked as farmworkers or other agriculturally related activities. Only 5 percent of the female population declared themselves to be unemployed.

When asked about their previous and current occupations, approximately one third of the women interviewed responded that they had had activities related to agriculture. About half of these women still hold jobs in agriculture or in some agriculturally-related activity. The decline in women's participation in agriculture was mirrored in other employment sectors such as industry. Formerly, 63 percent of women used to work in non-agriculture related activities in comparison to the current 42 percent working in that sector. This represents a decrease of 21 percent. In contrast, over time, the percentage of housewives increased from 11 percent to 37 percent, meaning that some formerly working women belonging to the *ejido* had become housewives. Because unemployed women were among the older group (the mean age of my sample was 50.7 years old), it can be affirmed that a generational difference existed

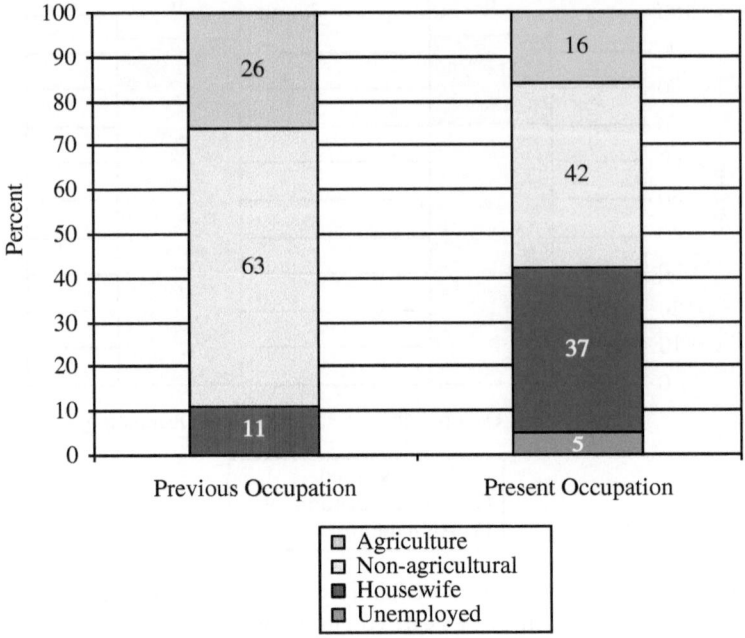

Figure 5.7 Previous and Present Occupations of the Female Labor Force in SLRC, 1994. *Source:* Structured interviews carried out in the *ejido*.

between those women holding jobs and those unemployed. It could also mean that women of 65 or older involuntarily became housewives or voluntarily retired. Women in SLRC, after a certain age, did not have the option to maintain their income-generating activities because many of them had worked in industry, or service, or commercial agriculture that required young female workers and turned away women after they reached a certain age.

Although the interviewees in the SLRC *ejido* had never met their grandparents, they stated that most of them, both their grandfathers and grandmothers, had been occupied in agriculture or some agriculture-related activity. The data gathered indicated that most inhabitants in SLRC were émigrés from other rural Western areas in which agriculture was their principal activity. Apparently for men, no significant occupational changes occurred since they continued working in agriculture either in Mexico or in the United States as shown by Figure 5.6. However, the data for women points to a change of occupations, especially for those who, before migrating to SLRC, worked in agriculture and, once in SLRC, worked in the *maquiladora* sector or in the services industry. The female labor force adapted to the labor demands of the economic activity located in this region.

The rates of immigration in SLRC were very high. These were confirmed first by the accelerated urban growth of the *ejido*. In addition, immigration patterns were tracked down by asking the birthplaces of the interviewees as well as those of their parents and grandparents. It was found that 56 percent of the *ejidatarios* came mostly from other Northwestern states such as Baja California, Sinaloa, Durango, Zacatecas, Nayarit, Jalisco, and Michoacán and 44 percent were from SLRC. Of the women *ejidatarias* only 31 percent had been born in San Luis Río Colorado and 16 percent in other parts of Sonora. In total, 47 percent of them belonged to the state of Sonora. Most women—about 53 percent—were immigrants from other Northwestern states such as Nayarit, Sinaloa, Michoacán, and Durango. The maternal and paternal origin of the interviewees corroborated the migration patterns in the area.

Transnational migration in this area proved detrimental for women and families in an *ejido* in which the average number of children a woman gave birth to was 5.2.[39] Families migrated into SLRC in order to cross the U.S. border. Once there, some of those men who were able to cross the border into the United States abandoned their partners and children in San Luis Río Colorado. In SLRC, the rate of woman-heads-of-household was high in comparison to Ixtaltepec, Oaxaca. About 58 percent of the women belonging to the *ejido* were supporting their families by themselves. Of these, 46 percent were single mothers, 36 percent were widows, and 18 percent were single. The remaining 42 percent were married women. Of these 26 percent declared that they also contributed to the family income. A 64 year old woman said in relation to women's work: "Too bad if [waged] women's work is not well viewed, anyway needy people have to work." However, another woman said: "Time ago, it was not common to see [waged] working women." They were referring obviously to salaried women, since salaried or not, women had always worked in Mexico.

LAND OWNERSHIP AND GENDER

In comparison to the other two *ejidos* of this research, in the SLRC *ejido*, more women owned *ejido* land. Evidently, these data refer exclusively to the members of the *ejido*, not to the total number of land owners in SLRC. About 17 percent of women were *ejido* owners in San Luis but different land ownership patterns existed. For example, some men and women residing in the SLRC *ejido* owned *ejido* land in other nearby *ejidos*, but not in SLRC. Some others were *avecindados* in SLRC. Usually *avecindados* and their families were recent arrivals in the *ejido*, they might own or rent their home but they did not hold the status of *ejidatarios*.

Of women owning land, 36 percent were single mothers, 29 percent were widows, 21 percent were married and 14 percent were unmarried. All the women who did not own land, about 26 percent of my interviewees, were married. In this *ejido*, there were two ways to gain access to *ejido* land, one through land grants and the other through succession (inheritance). About 71 percent of women owners of *ejido* land had received it through inheritance, while 29 percent of them had acquired it through a land grant. All the women who had been granted land were single mothers. When asked how they got their land, one of them told me that a friend had told her that people were organizing an *ejido* and she requested her name to be included. Some other women told me similar stories regarding their inclusion in the *ejido*. Both men and women residing in SLRC owned *ejido* land even if this was not in the SLRC *ejido*.

AVECINDADOS IN SLRC

Avecindados in SLRC owned *ejido* land in nearby *ejidos* such as El Fronterizo and La Grullita (see Map 5.2). Those *ejidos* were formed about 25 years ago. Some of the interviewees indicated that the organizers of those *ejidos* had invited men and women in need of a place or of a source of financial security to join. In the case of women, most of those invited were single mothers.

Therefore, the population initiating the formation of the *ejidos* adjacent to SLRC was not *ejidatarios* in the sense that they were devoted to agricultural activities in the area. They had not requested the land in order to work it. Rather, they were immigrants, mainly from other Western states, who had moved to Sonora and wanted to secure a piece of land for housing. By definition, they were *ejidatarios*, but by employment they moved through different identities working as farmworkers in the United States or as paid labor in other *ejidos*, in the case of men, or as *maquiladora* workers and housewives in the case of women.

Avecindados also worked in other economic sectors. They lived in SLRC as *avecindados* "because of the proximity to work or school," many of them had been renting their land for several years although before 1992 it was not legal to do so. *Avecindados*, in urbanizing *ejidos* as SLRC, are gradually outnumbering *ejidatarios*, and their role and influence need to be further explored as urbanizing *ejidos* transform.

DO YOU KNOW ABOUT PROCEDE?

When asked whether they knew about the 1992 Article 27, 79 percent of the interviewees responded that they did not know anything about it. The 21 percent who knew about it could not tell me what was new about this reform. Although PROCEDE had already been implemented in SLRC and the *ejido* was exhibited as the successful example of privatization and enterprise, only 5 percent of the *ejidatarios* knew when the program had been carried out and what it represented. When one of the interviewees was asked in the *ejido* office, during the very PROCEDE meeting to award her *ejido* land titles and certificates, whether she knew what PROCEDE was, she asked me: "Doesn't PROCEDE come from *proceder*?" She was right, *procede* comes from the Spanish verb "to proceed" or *proceder*, however, she did not know that I was referring to the PROCEDE program through which she was about to receive her land title. I ended up explaining to her the objectives of PROCEDE. Right in the middle of the privatization of *ejido* lands through PROCEDE, most of the *ejidatarios* were—at that time—unaware of the process and its consequences. It seemed that only the *ejido* authorities and some *ejidatarios* in SLRC knew about PROCEDE and its different options and alternatives. In spite of the lack of information to all the concerned *ejidatarios*, privatization had been carried out. In contrast, the *ejido* authorities in La Poza, Guerrero knew about PROCEDE and refused—at that time—to participate. The next chapter analyses the case of the *ejido* La Poza in the state of Guerrero.

Chapter Six

Privatization through Expropriation: The Case of the *Ejido* La Poza, Guerrero

When the *ejido* authorities of La Poza learned about the *Ejido* Reforms and PROCEDE, supposedly they refused to fully participate in the program and kept PROCEDE staff away from their lands. Because of the violence that different tourist agencies had periodically inflicted on the *ejido* population of La Poza, its residents distrusted everybody who approached the *ejido*. National and international capital, in its search for investment spaces, had colluded with local tourism development agencies which, in turn, had carried out violent expropriations of *ejido* land. These agencies took *ejido* lands, privatized them, and converted them to tourist purposes. Real estate developers consider precious the coastal lands of the *ejidos* in this region. The residents of La Poza had hoped not to attract the attention of tourist resort capital, although the tourism industry had a long history of relocation in this region. In fact, the history of the coastal region of Acapulco, which is where the *ejido* La Poza is located, is shared by other similar tourist regions in Mexico (see Map 6.1). They have experienced the same process of development at the cost of expropriating *ejido* lands and displacing their populations. They have gone through different privatizations, disguised as expropriations initiated by the State in the name of the public interest.

In this region, privatization of *ejido* land disrupted the internal economy of the *ejido* and its local labor force. The tourism industry that has occupied the space of the expropriated *ejido* has been unable to employ the labor force that its relocation displaced. This chapter presents the case of La Poza, its expropriation, privatization, and how the *ejido* reflects the relocation of capital, the expansion of Acapulco City, its effects in the region, and the continuous dispossession of Indigenous and local communities in the area during the time the 1992 *Ejido* Reforms and the PROCEDE were applied.

PACIFIC OCEAN

Map 6.1   Guerrero, Acapulco, and La Poza. *Source:* Drawn by author.

LOCATION AND DESCRIPTION

The *ejido* La Poza, also known as La Zanja,[1] is a coastal *ejido* in the Centralwestern state of Guerrero. It is located 45 minutes by bus South of Acapulco City and 8 minutes south of the *Diamante* tourist section of Acapulco (Map 6.2).

Like San Luis Río Colorado, La Poza is an *ejido* rapidly transforming into an urban area. In the case of La Poza, this transformation has taken place because of the *ejido's* proximity to a rapid-growing tourist city like Acapulco. For the low-income populations in-migrating to this area, *ejido* land for housing purposes has traditionally been more affordable to rent or acquire than land in the city. La Poza used to be an *ejido* devoted to agriculture and fishery. The *ejidatarios* have also tried raising cattle, coconut oil extraction, and lately, due to land conversion and the subsequent reduction of agricultural land, plant nurseries. The development of the tourism industry in the region has violently displaced and relocated an *ejido* population[2] that consists of Blacks and Indigenous groups of the region. Historically, coastal *ejido* land has been expropriated for tourism purposes and subsequent reorganizations of land and labor have occurred.

Acapulco is both a municipality and a city comprising 42 *ejidos* and agrarian communities, the *ejido* La Poza is one of them. According to the official census, in 1990, the city of Acapulco had approximately 515,000 people.[3] Several city officials, however, warned me that the census had severely undercounted the city's population which, by the end of 1995, they estimated to be more than one million.[4] Officially, the population of *ejidatarios* in the municipality was estimated at 5,150.

Before 1992, the census did not provide data for every single *ejido*. It was possible to obtain a rough estimate of the population in La Poza because

ACAPULCO CITY

TRES PALOS   SAN PEDRO
                    DE LAS PLAYAS

ACAPULCO
BAY   TRES PALOS
                             LAGOON

PACIFIC   LA
OCEAN   POZA

N

Map 6.2   The *Ejido* La Poza. *Source:* Instituto Nacional de Estadística, Geografía e Informática (INEGI), *Acapulco de Juárez, Estado de Guerrero: Cuaderno Municipal, Edición 1993* (Aguascalientes, México: INEGI, 1994), 10. Based on CGSNEGI, Carta Hidrológica Aguas Superficiales.

a medical practitioner working for a clinic in the *ejido* had undertaken a local census in 1993. He had counted 604 men and 577 women. This census, however, only considered those who received benefits from the clinic and counted both *ejidatarios* and *avecindados*.[5] Therefore, we can assume that the total population was above the 1,181 people included in that local count.

Although the local census did not specify the composition of the population, the population of the coastal area of Guerrero has been predominantly either of Indigenous origin or of African descent. The Acapulco population has often been recognized as a mix of Indigenous and Black

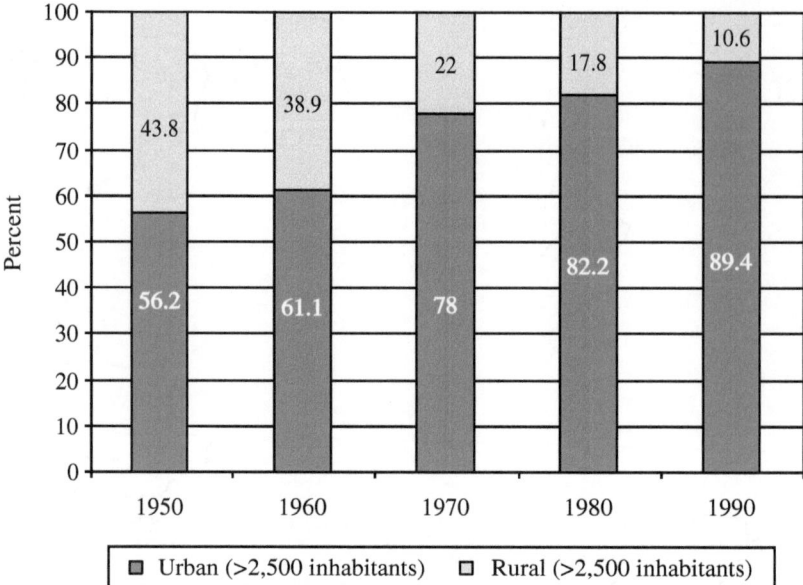

Figure 6.1   Urban and Rural Population in Acapulco, 1950–1990. *Source:* Instituto Nacional de Estadística, Geografia e Informática (INEGI), *Acapulco de Juárez, Estado de Guerrero: Cuaderno Municipal, Edición 1993* (Aguascalientes, México: INEGI, 1994), 16: Based on data from INEGI, "Guerrero, Resultados Definitivos. VII–XI Censos Generals de Población y Vivienda, 1990."

people.[6] Guerrero houses more than seven Indigenous groups in its territory. According to the 1990 census data, at the state level 11.4 percent of the population speaks an Indigenous language and, in the municipality of Acapulco, only 2 percent of the population speaks an Indigenous language.

My observations both in the city and in the *ejido* indicate that if anybody had been undercounted in the census it was the Indigenous and Black populations at the state, the municipal, and the *ejido* levels. The populations of the *ejido* were mostly of Indigenous and African descent and in Acapulco city, Indigenous people worked in the streets, on the beach and in the formal and informal market places.

In addition, the census indicates that the rural population in Guerrero has dramatically decreased since 1950. In that decade, the urban population accounted for 56.2 percent while the rural population represented 43.8 percent. In the latest count in 1990, the urban population was 89.4 percent, while the rural population was only 10.6 percent (see Figure 6.1). However, this count radically differed from my observations and data gathering in the *ejido*, which made me believe that the rural population is also severely undercounted. In spite of the primacy of Acapulco in the state of Guerrero, the state continues to be a predominantly rural and Indigenous region.

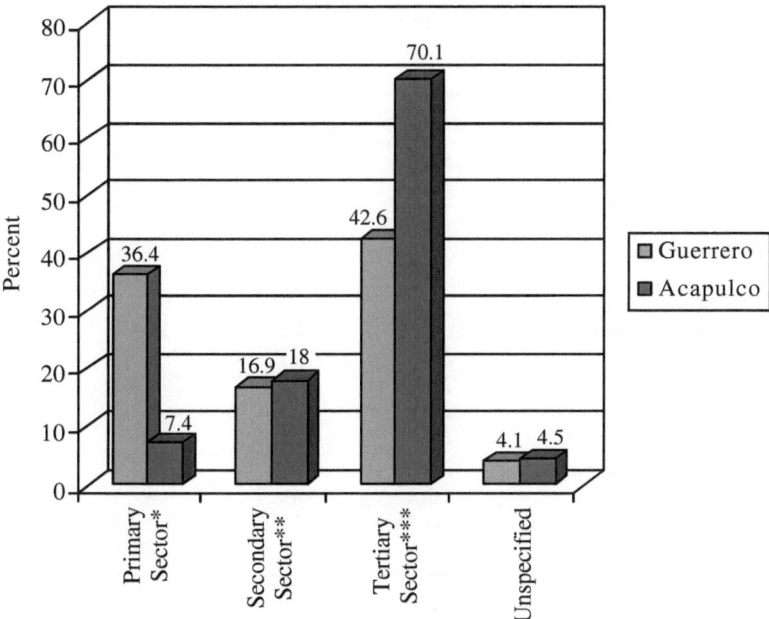

Figure 6.2 Employed Population by Economic Sector in Guerrero and Acapulco, 1990. *Includes agriculture, livestock, forestry, and fishing. **Includes mining, oil and gas extraction, manufacturing industry, generation of electric energy, and construction. *Source:* Instituto Nacional de Estadística, Geografía e Informática (INEGI), *Acapulco de Juárez, Estado de Guerrero. Cuaderno Estadístico Municipal. Edición 1993* (Aguascalientes, México: INEGI, 1994), 60. Based on data from INEGI, "Guerrero, Resultados Definitivos. XI Censo General de Población y Vivienda, 1990."

THE ECONOMY OF THE REGION

Tourism is the main economic activity in the region along with agriculture, fishing, and services. Coastal and inland *ejidos*, in close proximity to Acapulco service the different needs of the tourism industry. The dominance of employment in the tourism industry is evident even at the state level. Figure 6.2 shows that in 1991, at the state level, 42.6 percent of the population were employed in the tertiary sector, while employment in the primary sector accounted for 36.4 percent, and in the secondary sector amounted to 16.9 percent.

In comparison, the municipal data showed that employment in the tertiary sector in Acapulco encompassed 70.1 percent of the total municipal economy, while the secondary sector employed 18 percent of the population and only 7.4 percent was employed by the primary sector.

TABLE 6.1   Employment Indicators in the Metropolitan Area of Acapulco, 1992

	Trimester			
Indicator	I	II	III	IV
Employment	52.2	52.9	53.7	52.8
Men	72.6	73.1	72.5	71.5
Women	34.1	35.0	36.7	36.0
Unemployment*	2.9	2.3	1.7	0.9
Men	2.8	2.5	1.6	0.9
Women	3.2	2.0	1.9	0.8
Reason of Unemployment	100	100	100	100
Termination	29.6	37.2	20.1	44.5
Temporal Job	26.0	26.0	40.9	26.4
Unhappy with Job	21.8	15.4	13.1	11.2
Other	22.6	21.4	25.9	17.9
Duration of Employment	100	100	100	100
1–4 weeks	60.3	49.5	47.1	42.6
5–8 weeks	20.2	11.6	24.4	29.3
9 or more weeks	19.5	38.9	28.5	28.1

*Population over 12 years old unemployed for a period of 2 months at the time of the [census] interview. *Source:* INEGI, "Cuaderno de Información Oportuna," No. 244 (México: INEGI: Julio 1993)" as part of the table of the Instituto Nacional de Estadística, Geografía e Informática (INEGI), *Acapulco de Juárez, Estado de Guerrero: Cuaderno Municipal, Edición 1993* (Aguascalientes, México: INEGI, 1994), 61.

The instability and temporality of the jobs offered by the tourist industry are characteristics well represented in Table 6.1 which shows the reasons why labor becomes unemployed in the metropolitan area of Acapulco by trimester.

Because of the seasonality of tourism in the area, the fluctuations in the data correspond to the temporary hiring of personnel in the industry or in tourist-related employment areas and in their abrupt dismissal according to the needs of the tourist industry. The data show that at the end of the summer of 1992, the third trimester, 40.9 percent of the employed population lost their temporary jobs. By the end of the second and fourth semesters, respectively, 37.2 percent and 44.5 percent of those holding permanent jobs had lost them. The same table shows the short duration of employment offered in the region. The data for each semester shows at least 42.6 percent of the employment offered each semester lasted only from 1 to 4 weeks, while no more than 28.1 percent of the jobs lasted 9 weeks or more.

The seasonality of jobs translates into low wages. Wage statistics shown in Figure 6.3 indicate why Guerrero is considered one of the poorest states in Mexico. In the municipality of Acapulco, more than 27 percent of the population makes less than minimum wage or no wage at all, 37.9 percent earns less than twice the minimum wage. In general, at least 65 percent of

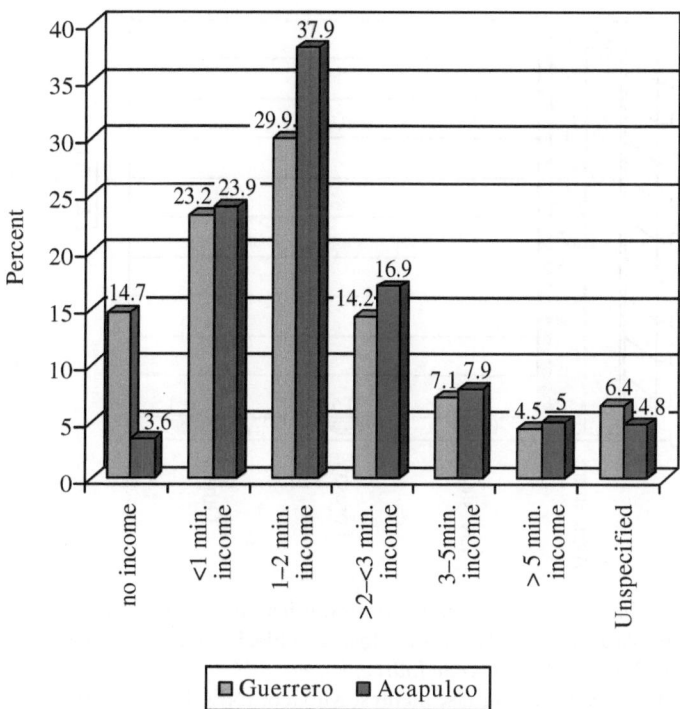

Figure 6.3 Employed Population by Monthly Income in Guerrero and Acapulco, 1990. *Source:* Instituto Nacional de Estadística, Geografía e Informática (INEGI), *Acapulco de Juárez, Estado de Guerrero. Cuaderno Estadístico Municipal. Edición 1993* (Aguascalientes, México: INEGI, 1994), 61. Based on data from INEGI, "Guerrero, Resultados Definitivos. XI Censo General de Población y Vivienda, 1990."

the population of the municipality of Acapulco is either low income or no income as opposed to the previous case of San Luis Río Colorado in which only 43.3 percent were low-income.

THE ECONOMY OF THE *EJIDOS*

The *ejidos* and agrarian communities both in the state of Guerrero and in the municipality of Acapulco were devoted mainly to agriculture or agriculture-related activities.

Figure 6.4 shows that in 1991, 97.6 percent of the *ejidos* and agrarian communities in the municipality worked in agriculture, while 2.4 percent worked in other non-specified activities. Although fishing and livestock were economic activities undertaken both in the municipality and in the *ejido*, the census figures indicated as "confidential" the data on these sectors,

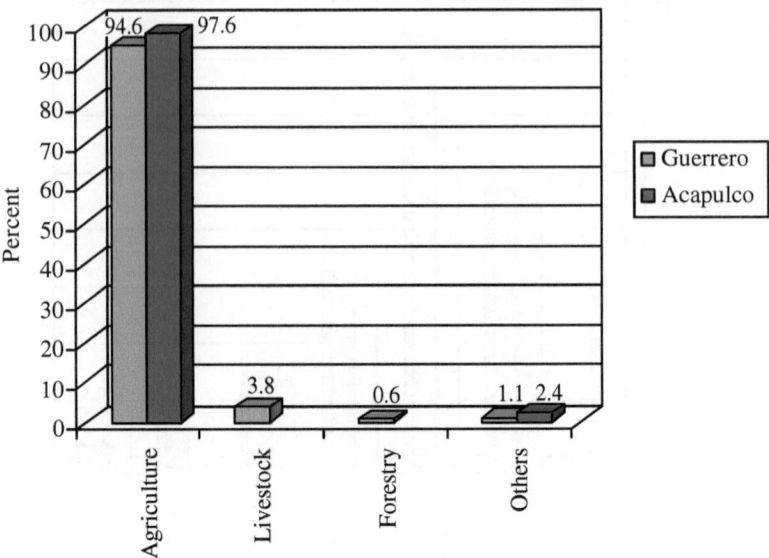

Figure 6.4 *Ejidos* and Agrarian Communities by Main Economic Activity in Guerrero and Acapulco, 1991. *Source:* Instituto Nacional de Estadística, Geografía e Informática (INEGI), *Acapulco de Juárez, Estado de Guerrero. Cuaderno Estadístico Municipal. Edición 1993* (Aguascalientes, México: INEGI, 1994), 68. Based on data from INEGI, "Guerrero, Resultados Definitivos. VII Censo Ejidal, 1991."

and instead of numbers, the letters ND, which stood for *no disponible* or "not available," appeared.

Data provided by the city government specifies land uses in the municipality and its *ejidos*. According to that data, the municipality comprises 188,260 hectares. Of these, 18 percent is devoted to agriculture, 14 percent to livestock, 55 percent to forestry, and 13 percent to other uses (see Figure 6.5).

Figure 6.6 shows that from the total of hectares, 69 percent were *ejido* lands; while 20 percent belonged to agrarian communities, 6 percent were considered private property and the remaining 5 percent were federal public lands. The total number of *ejidatarios* in the municipality was 5,150. Since most of the Acapulco land had been designated *ejido* land, it is fair to infer that Acapulco, like SLRC, had followed the same urbanization pattern as that of Mexico City, meaning that it had grown on *ejido* land.

ACAPULCO CITY

The tourist area of Acapulco is divided into three main sectors: the Traditional Acapulco, the Golden Acapulco, and The Diamond Acapulco.[7] These sectors represent different urbanization and tourist development stages of the city. Traditional Acapulco is the original part of the city that

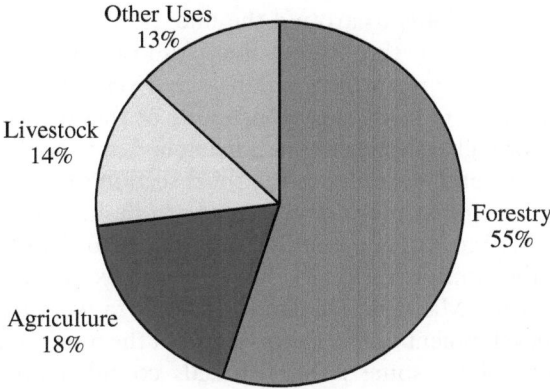

Figure 6.5 Land Use in *Ejidos* and Agrarian Communities by Main Economic Activity in Acapulco, 1993. *Source:* Data provided by the Planning Office of the City of Acapulco.

developed from colonial times on. Some of the remaining architecture in this section of the city reveals its past as one of the main city-ports in Mexico. Currently, its hotels and neighborhoods are considered the most affordable for national tourists and for low-middle class residents of the city. Traditional Acapulco is a mixed-use area, with affordable hotels, public offices, commercial areas, and public parks. It also encompasses downtown Acapulco. The second tourist section, Acapulco *Dorado* or Golden Acapulco, was

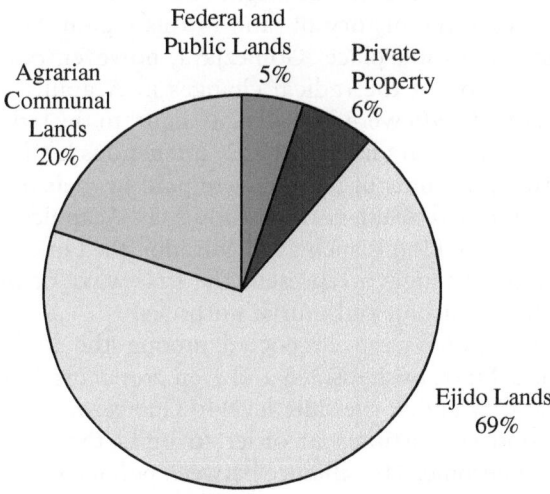

Figure 6.6 Land Tenure in Acapulco, 1993. *Source:* Data provided by the Planning Office of the City of Acapulco.

developed from the 1940s to around the mid 1980s. It is located South of Traditional Acapulco on the coastal line. Almost no average resident of Acapulco lives in this area which is considered very expensive. Generally, condos and apartments in the neighborhoods of this section are rented to foreign tourists or high-income nationals living in Acapulco. Finally, Acapulco *Diamante* or Diamond Acapulco is the third section of tourist Acapulco. It was planned as a tourist mega-development to be built on approximately 76,000 hectares.[8] Most of the land was coastal *ejido* land that was expropriated under the administration of the murdered ex-governor of Guerrero José Francisco Ruiz Massieu. The lands of La Poza were partially affected as part of the development of this third section of the tourist city of Acapulco. During my interviews, some public officials confidentially declared that land for the Punta Diamante project was expropriated for the benefit of politicians involved in the mega-project. The conflict of interest present in San Luis Río Colorado with the *ejido* authorities also occurred in La Poza with politicians since many of them were direct or indirect beneficiaries of the *ejido* land expropriation.

ILLEGAL PRIVATIZATION THROUGH ILLEGAL EXPROPRIATION

Expropriations in Acapulco are a commonly used instrument to privatize *ejido* land for all kinds of illegal purposes. Francisco Gomezjara has documented the history of privatizations and the dispossession of Indigenous lands in the region and the violence that erupted because of those illegal expropriations.[9] This violence is considered a current defining characteristic of the region. In a way, to say "Guerrero" is to say "violence."

But violence in Guerrero has an origin and, in order to understand it, it is necessary to learn the history of land in this region. Since times of the colony dispossession took place. Gomezjara, however, points to the year 1928 as the initiation of the radical changes in Acapulco.[10] In that year, lands of the colonial *ejido* were awarded, as a gift, to the richest commercial men in the area. Four years later, in 1932, Indigenous lands were expropriated and microscopic sums of money were paid to their owners. The area of this dispossession is what is now known as Acapulco *Tradicional* or Traditional Acapulco. Hotels such as El Mirador, the famous La Quebrada, Las Hamacas, and Papagayo represent the first wave of dispossession of Indigenous lands for urban and tourist purposes.

Such expropriations were concocted among the highest politicians, members of the cabinet, businessmen and even presidents, both at the federal level in Mexico City and at the state level in Guerrero.[11] They even created joint construction consortiums in order to build the tourist Mecca that Acapulco would become. This alliance between politicians, military officials, and business people to illegally obtain land is indicative of the character of the Mexican State that was described in Chapter Two. There was in this instance no distinction between capital and the State.

The history of tourist Acapulco then, is the history of the dispossession of Indigenous lands. In 1940, for example, the Icacos *ejido* started to be gradually expropriated and then privatized. The original public-good reasons for the expropriation were, in most cases, transformed and the land ended up in private hands. Promised payments to the owners of the *ejido* were never fulfilled. In 1960, in the greed for privatizing and developing this precious coastal land, even the *ejido* cemetery was bulldozed. The Holiday Inn Hotel and several luxurious condos were built on that site.[12] By 1945 new expropriations took place in what is known now as the Caleta and Caletilla beaches. Street vendors and merchants were violently removed by the army because they "made ugly that modernizing zone."[13] At the same time, in Puerto Marqués, *ejidatarios* and their families were violently displaced from the *ejido* and some of them were killed by the army. In 1950, the *ejido* Cumbres de Llano Largo was violently dispossessed and the *ejidal* leader assassinated. In 1970, it was the turn of the Barra Vieja *ejido*. Here, the state government jailed all the male heads of households of the 40 families of the *ejido*, so that they would agree to sell their lands.[14] Finally, in 1980, expropriations to build Diamond Acapulco started when the Tres Vidas resort, financed by foreign capital, would dispossess the inhabitants of the Plan de los Amates *ejido*.[15] Other *ejidos* followed the expropriation path as Mozimba and the *ejido* La Zanja, that is now known as La Poza, started to be gradually displaced.

IN THE NAME OF THE PUBLIC UTILITY

Expropriations of *ejido* land have been executed in the name of "public utility" and against the will of the affected *ejidatarios*. They have generated extreme violence in the state officially considered as the poorest in Mexico. *Ejidatarios* opposing expropriation of their *ejidos* have been murdered while others have been sent to jail. Those who "accepted" expropriation have been paid minimal prices for their land, while the government sold expropriated *ejido* land to foreign investors at very profitable prices. In recent expropriations, *ejidatarios* were paid between one and three dollars per square meter for their land. The same land was resold to investors at prices ranging from $70 to $150 per square meter. In a project called Real Diamante, land fetched prices between $1,500 and $2,300 per square meter.[16]

Through expropriations, national and foreign investors have gradually concentrated land. In order to carry out these expropriations, the government of the state of Guerrero created *Promotora Turística* (PROTUR) or Tourism Promoter Agency that was defined as a "decentralized agency." After expropriating *ejido* land, PROTUR offered it to potential foreign investors. In one of its informational reports, PROTUR described the expropriated land as "Territorial Reserve" and offered it "directly [. . .] for tourism development, whether they [the lands] are State or privately owned."[17] The land use planning objective of the Territorial Reserves became into a speculative motive in Acapulco. In this area, Territorial Reserves were used as a pretext to dispossess local communities, destroy their housing and their local economies, and offer their lands to national and international investors.[18]

EXPROPRIATIONS IN LA POZA

The *ejido* La Poza has gone through several transformations because the government expropriated its land and relocated its residents. Some of them were moved to a hilly area with no services called Miramar del Marqués. Others were moved to a strip of land behind the original *ejido* and next to the Tres Palos Lagoon. In the near future, developers plan to attain the *ejido* land around the lagoon as well as the lagoon itself.[19] This will involve additional expropriations of *ejido* land and consequently, a second relocation of the original residents of La Poza as well as other affected *ejidos*.

According to the *Comisariado Ejidal*, La Poza was founded in 1935. Previously, it had been a large *latifundio* called Hacienda El Potrero which encompassed not only La Poza, but also other current *ejidos*, including Puerto Marqués. Hacienda El Potrero was owned by the Stephen family who were livestock farmers, originally from the United States. One of the Stephen family's contractors was the father of the current *Comisariado Ejidal* in La Poza and was one of the founders of the *ejido*. His son remembers how his father traveled to Tecpan, Guerrero to bring his brothers to La Poza in order to gather the required number of people to apply for an *ejido* land grant. Originally, the *ejido* was named La Zanja, but later became known as La Poza because the *ejido* had a large well. Unfortunately, the well eventually became extremely polluted because nearby *ejidos* dumped their drainage in the well.

The *ejido* lands of La Poza were gradually decimated. The first expropriation occurred in 1945 during the administration of Miguel Alemán. A second taking occurred in the late 1970s. Finally, land was re-expropriated in 1987 by Ruiz Massieu. Some of the previous owners have still not been compensated.[20] The resorts built in this area enjoy a privileged location because of their proximity to the newly constructed *Carretera del Sol*[21] and to the Acapulco airport. As *ejidos* in this region lost land, tourist capital gained space to invest in tourist resorts.

VIDAFEL OR HAPPY LIFE

In lands of the *ejido* La Poza, the Vidafel resort was built. Vidafel stands for *Vida Feliz* or Happy Life in Spanish. As soon as Vidafel started operations, it closed the traditional accesses to the beach by building a 4-kilometer wall all around the resort. The wall impedes public access to the beach for both *ejido* population and for regular visitors, who now have to walk or drive about 3 kilometers to find a public access to the beach. Since the construction of the wall and the resort's development, people have been banned from engaging in any economic or recreational activity on that beach. Agriculture was no longer an option. In addition, fishermen from the *ejido* were banned from fishing along the beach and women were prohibited from selling produce or jewelry. Some *ejidatarios* and fishermen who tried to tear apart the wall are currently in jail for damaging private property. This wall separates the disturbing luxury of Vidafel from the modest houses and nonpaved roads of what was left of the *ejido*. It separates two different spaces and two different groups of users of that space: tourists and *ejidatarios*.

THE RESTRUCTURING OF THE LOCAL ECONOMY

The tourism industry has had an irreversible impact on land and labor in the region as *ejido* land was expropriated and traditional jobs disappeared. *Ejidatarios* in La Poza used to work as farmers and fishermen. Over time and by means of expropriations, they moved from agriculture to coconut production and coconut oil extraction. They also participated in government programs to produce livestock. They fell into debt because those programs failed and impoverished them further. They continued to identify themselves as *ejidatarios* although some of them not only in La Poza, but also in other *ejidos*, had become carpenters, construction workers, taxi drivers, gardeners and janitors, maids, cooks, and street vendors. Their occupations changed according to the seasonality of tourism and its demands.

In spite of the dependence of labor markets on tourism, Acapulco was considered in the early 1990s as a resort area-in-decline.[22] The tourism industry was destroying local jobs and was not generating new ones. In addition, the pollution of Acapulco Bay and the environmental degradation in the region had decreased the numbers of national and international tourists visiting the area. Although the owners of the Vidafel project had assured everyone that the development would create about 1,000 direct jobs, when I visited the resort, it was evident that none of the local residents of the *ejido* were working at Vidafel. Thus, if jobs were created, they were not for the displaced fishermen, farmers, or women selling jewelry on the beach.

When I first interviewed the *Comisariado Ejidal*, he suggested that I not stay in the *ejido* because I did not know anybody. I did not regard this as a serious motive, so I tried to find accommodations. I did not succeed. Because of the long history of dispossession and violence in the region, it was evident that residents of the *ejido* did not trust me. Nobody could tell me of a place to stay. Thus, I ended up commuting everyday from Acapulco to the *ejido* La Poza. This daily commute gave me an insight into the kind of population that commutes from the city to nearby *ejidos* and vice versa. It also made me aware of the fact that La Poza is rapidly becoming part of Acapulco. I observed that several women, some of them wearing uniforms like those of domestic workers, would get off at the Golden and the Diamond areas of Acapulco. Evidently, they worked for hotels and resorts on the coast.

THE INTERNAL FISSURES WITHIN THE *EJIDO*

La Poza, like San Luis Río Colorado, had a history of internal fissures and conflicts. The main contention was the allegation of *ejidatarios* accusing some members of the *ejido* authorities, claiming that both the previous and the current *Comisariado Ejidal* had negotiated under the table with PROTUR and other governmental agencies, in order to secure individual profit from *ejido* land expropriations. Another point of dissension was the subdivision and sale of *ejido* land by *ejidatarios* who, either because of financial need or fearing further expropriations had decided to sell part of their land. The people in the group buying that *ejido* land were called *avecindados* and they did not acquire *ejido* rights by buying land within the *ejido*. In addition, in many cases the sale was illegal.

According to *ejido* documents, La Poza was composed of approximately 32 *ejidatarios*, 10 of them women, and 203 *avecindados*.[23] *Ejidatarios* were the only ones who could determine the fate of the *ejido*. Although *avecindados* outnumbered *ejidatarios*, they did not have the right to speak out or to vote in *ejido* assemblies. Decisions to enter the PROCEDE to title the *ejido* land or to accept new *ejido* members or *avecindados* were in the hands of those 32 *ejidatarios*.

THE INFORMAL SALE OF *EJIDO* LAND

The *Comisariado Ejidal* told me that although official documents reported 32 *ejidatarios*, there were actually 46 *ejidatarios*. He explained that 12 *avecindados* had applied to be converted into *ejidatarios* and their request had been accepted by the *ejido*. The *Comisariado* did not tell me the mechanisms they followed to convert *avecindados* into *ejidatarios*. However, one morning, while I was reviewing documents at the *Comisariado Ejidal* house, a white, female *avecindada* visited the *Comisariado*. She expressly asked him to advocate for her acceptance as an *ejidataria* in the next *ejido* meeting. In addition, she asked the *Comisariado Ejidal* for his signature to validate an *ejido* land sale she was about to carry out. Normally, the *Ejido* Assembly is the proper forum to request this kind of validation or to decide who will become *ejidatario*. Apparently, ignoring the decision-making authority of the *Ejido* Assembly was a common practice in La Poza *ejido*.

The woman was well dressed, lived in Acapulco and, for several years had been buying land in La Poza. During the conversation she emphasized "all the favors I have done for you and Fermina" (the previous *Comisariada Ejidal*) as a basis for positively considering her requests. Since the woman did not know who I was, she did not care about my inoffensive presence, nor did the *Comisariado Ejidal*. At this point, the *Comisariado Ejidal* responded that he would bring the subject up at the next *ejido* meeting. If the *ejido* members accepted her request, she would need to give a contribution "to help the *ejido*." The woman stressed that she could not pay any compensation to the *ejido* for selling her plot because she was poor and could not afford it. Elegantly, she left her business card and asked to be called in case he needed "any favor or to arrange anything." I had witnessed a negotiation process of *ejido* land and change of status within the *ejido*. The "contribution" was a bribe asked in order to achieve the status of *ejidataria* and to allow her to sell one of her several *ejido* land plots. It was evident that the *Comisariado Ejidal* was accustomed to carrying out *ejido* land sales without the legal consent of the *Ejido* Assembly.

A half and hour later, an old Indigenous *ejidataria* woman came in, accompanied by a man who was going to buy her land. She had had a family emergency and was in need of the money. By this time, the *Comisariado Ejidal* was convinced that I was completely inoffensive or that I did not understand what was going on. In any case, I was in his territory. He renewed the ritual and asked again for a "cooperation" to legalize the sale. He asked the equivalent amount of money that she would obtain from selling the plot. She was selling it for one million pesos[24] and the *Comisariado Ejidal* was requesting a similar amount of money from her. Rigidly, he remarked that in case she did not comply with that "cooperation," he would not make the sale legal. The difference between this woman and the Acapulco resident with *avecindada* status was striking. It was evident that this older woman was very low-income, of Indigenous origin, wearing her *rebozo*[25] covering her head and in addition she was a real resident of the *ejido* while the other woman was not. The other woman, in connivance with the *Comisariado Ejidal*, was participating in the profitable business of speculating with *ejido* lands. Thus, in addition to the pressures of national and transnational capital for privatizing coastal *ejido* lands, the internal relations in the *ejido* worked, at a different level, for that privatization.

At this point, and spite of my critical view of the *Ejido* Reforms, I recognized the advantage for *ejidatarios* and *avecindados* of having their own certificate or title for their land, at least for *ejidos* of this kind. Not having to depend on corrupt *ejido* authorities to validate any single operation within the *ejido* meant taking away power from dishonest authorities that, in many cases, had become political and economic bosses. In any case, with or without a certificate, *ejidatarios* and their families were pushed by economic and political reasons to sell their land or they faced expropriation of their lands in the name of "the public interest."

LAND PRIVATIZATION EQUALS LAND DEPRIVATION

The generalized context of violent expropriations and the internal fissures within the *ejido* gave me a different perspective of the regional differences and the unevenness of development and social justice in *ejidos*. During field research I had to face the effects of the violence undergone by *ejidatarios* and their families. For example, in order to facilitate my work, I requested from the *Comisariado Ejidal* a presentation letter to introduce myself and the objectives of my work to *ejidatarios*. However, every time I extended the letter, people's distrust increased. A woman even threw me out of her place believing I was a PROTUR agent. She violently shouted at me, nodding and yelling: "Now I understand. You came here to deprive us of our land. Your letter here says that you are doing a study on Land Deprivation." I tried to explain that the letter said Land Privatization,[26] not Land Deprivation. She replied: "Well, land privatization and land deprivation is the same thing, isn't it?" I could have explained to her that they were different procedures, but I just couldn't. What *ejidatarios* in this region were experiencing matched the meaning of these two words and, in fact, land privatization had become synonymous with land deprivation for the inhabitants of this *ejido*.

When I mentioned the incident to other residents of the *ejido*, they explained that they had experienced the taking of their land, harassment by police and developers, arrests, and bulldozing among other things. They did not trust anybody anymore. When developers had sent somebody to ask about the limits of the *ejido*, residents naively had given that information to later find out that the person requesting that data was working for the developers. They had learned not to provide information to anybody.

Some *ejidatarios* told me that one of the reasons for their abhorrence of the *Ejido* Authorities was that the compensation for the last expropriation, officially paid to the *Comisariado Ejidal*, was never redistributed to the affected *ejidatarios*. The *Comisariado Ejidal* at that time was the widow of the previous *Comisariado Ejidal*. She had received the compensation from PROTUR but, according to my interviewees and to several newspaper articles in Acapulco,[27] she had kept most of the money and some *ejidatarios* had not received any compensation for their expropriated land. Thus, in addition to facing violent expropriation and displacement of their lands, *ejidatarios* had to relinquish the already minimal compensation for their lands.

VISITING VIDAFEL

Ejidatarios were not the only ones reluctant to talk about the *ejido*. Developers in charge of the project never gave me an interview. Public officials, when interviewed, simply emphasized the modernization process taking place in the "backward" areas of Acapulco. The modernization process basically meant the exclusion of Indigenous communities, *ejidatarios*, farmers, fishermen, women merchants, street vendors, and other "ugly" populations from their coastal lands.

I managed to visit the Vidafel resort by making an appointment with a salesperson from the development. Access to the resort had been restricted to visitors and workers in order to prevent further protest demonstrations from affected *ejidatarios*. A previous appointment was necessary because Vidafel was safeguarded like a fortress. Appointments were available only for those tourists interested in obtaining information to buy a time-share condo in Vidafel.

According to the salesperson, the investment channeled to develop Vidafel was the highest in Latin America for a resort of that type.[28] Because of its overwhelming luxury, many *ejidatarios* commented that "narcos"[29] had laundered their money building Vidafel. Nobody in the *ejido* nor in the city provided me with evidence to document that rumor, but residents of Acapulco agreed with *ejidatarios* and mentioned that tourism was one of the most profitable industries in which money could safely be cleaned up. If this was true the criminal economy had found a safe place to invest: the coastal *ejido* land.

Although no evidence of such investment was available, Vidafel exhibited an exaggerated ambiance of sumptuousness; marble floors, exotic gardens, fish ponds, real flamingoes and other decorative birds, and a pool at least 1,250 meters long. No tourist could ever imagine that the area used to be an *ejido* with modest homes, no pavement, and no urban infrastructure, nor were they aware that, in order to build such a fantasy, the government, in collusion with national developers and transnational capital, had violently displaced the residents of the *ejido* La Poza.

EJIDATARIOS, AVECINDADOS, AND SQUATTER SETTLERS

The population of the *ejido* La Poza is similar to that of the San Luis Río Colorado *ejido* because most of its inhabitants are not *ejidatarios*, but *avecindados*. Although La Poza is not a city, it has undergone a rapid land conversion because of its proximity to Acapulco. In the near future, La Poza will become part of the Metropolitan area of Acapulco or will be completely expropriated to build more tourist resorts. The *Comisariado Ejidal* did not provide me with a list of *ejidatarios* and their addresses, although he did tell me that of the approximately 250 heads of households, 80 percent were *avecindados*. With no specific addresses at hand, I decided to conduct random interviews going home by home, since the *ejido* was not as big and did not have the same sprawling layout as SLRC. The disadvantage of choosing this random interviewing approach for La Poza *ejido* was that I was not going to interview the "real" *ejidatarios*, but both *ejidatarios* and *avecindados*.

To my surprise, I also met and interviewed squatters who had taken over a piece of the *ejido* since 1992. *Ejido* residents explained that, previous to the arrival of the squatters, a *"narco"* had illegally appropriated that land. The squatter invasion occurred after he was denounced and ran away. The name of the squatter settlement was Colonia Las Delicias. Thus, *ejidatarios*, *avecindados*, and squatters were the three groups sharing the space of the *ejido* La Poza. The presence of these three divergent groups on the *ejido* is a manifestation of the transformation of the *ejido* and of the ongoing struggle of different groups for land.

As expropriations had taken land away from *ejidatarios*, they also took from them their economic support and their housing space. It can be hypothesized that in La Poza *ejido*, and probably in the entire Acapulco tourist region, a good number of the *avecindados* and squatters had come there as the result of violent expropriations and displacement from their original lands. In the same vein, rapid urbanization in the municipality of Acapulco has taken place not only due to in-migration, but also because displaced *ejido* populations have urbanized previously vacant land.

Within this context, the results of the interviews indicated that 28 percent of the interviewees declared themselves to be *ejidatarios*, 39 percent were *avecindados*, 22 percent did not own any land, meaning they were living in the *ejido* because they were renting either *ejido* or *avecindado* land and, finally, 11 percent were squatters. *Avecindados* were reluctant to provide information about the way they had obtained *ejido* land. Some of them mentioned that they "just bought it." *Ejido* land sale in La Poza had been so widespread that even *avecindados* had started to sell their lands. There were cases also in which *avecindados* did not live in the *ejido*, but rented out the land they had acquired. All kinds of illegal *ejido* land transactions were allowed, and internally profited from, by the *Comisariado Ejidal*. Thus, *ejidatarios* had to face dispossession and displacement from external forces and internally they had to deal with corrupt *ejido* authorities.

An interesting development within the *ejido* was the growing participation of female *avecindadas* who had started to attend *ejido* meetings in spite of not having the right to vote or to speak. They attended in their attempts to start organizing the *avecindado* population in order to request urban services and land regularization.

MIGRATION WITHIN THE *EJIDO*

In addition to the constant decimation of the lands of La Poza, migration into Acapulco and its surrounding area was also contributing to the transformation of the *ejido*. The results of the interviews indicated that 46 percent of the male residents of the *ejido* came from other rural areas in Guerrero, 20 percent had migrated from Acapulco, another 20 percent came from expropriated *ejidos*, 7 percent had come from the neighboring state of Oaxaca, and only the remaining 7 percent were natives of La Poza. Meanwhile, 28 percent of the female population had migrated to La Poza from other rural areas in Guerrero. Displaced female populations from nearby expropriated *ejidos* also amounted to 28 percent, migration from Acapulco City represented 22 percent and 5 percent had migrated from the state of Oaxaca. Only 17 percent had been born in the *ejido* La Poza.

These figures indicated no significant interstate migration, but a significant intrastate migration from other rural areas of Guerrero into the municipality of Acapulco. This migration might have been motivated by the generation of jobs in the tourism industry. Interestingly, however, just a few of my interviewees were directly employed by the tourist industry and, those who were, held only temporary jobs during peak vacation periods.

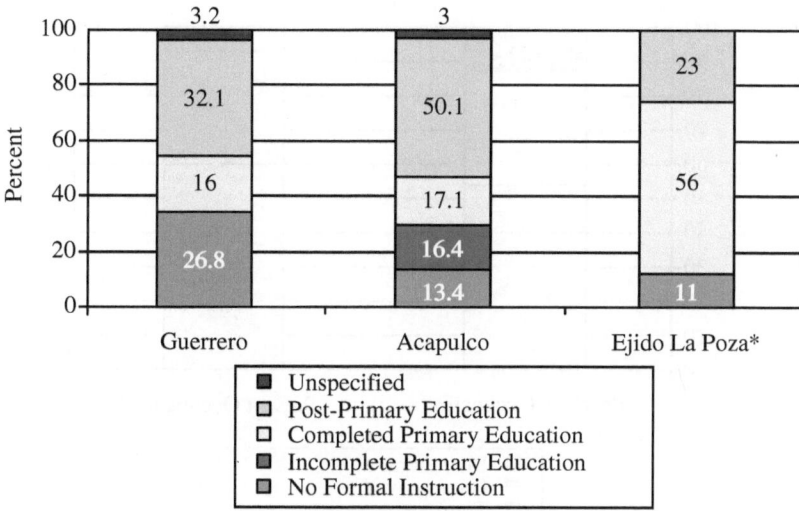

Figure 6.7 Population 15 Years and Older by Educational Level in Guerrero, Acapulco, and the *Ejido* La Poza, 1990. *Data obtained from the results of the structured interviews carried out in the *ejido* by author in 1994–1995. *Source:* Instituto Nacional de Estadística, Geografía e Informática (INEGI), *Acapulco de Juárez, Estado de Guerrero. Cuaderno Estadístico Municipal. Edición 1993* (Aguascalientes, México: INEGI, 1994), 44. Based on data from INEGI, "Guerrero, Resultados Definitivos. XI Censo General de Población y Vivienda, 1990."

EDUCATION AND JOBS IN THE TOURIST INDUSTRY

Educational levels in the *ejido* were very low (see Figure 6.7). Although at the state and municipal levels, the illiteracy rates were of 27.5 percent and 25.6 respectively, at the *ejido* level it was 11 percent. Approximately 56 percent of my interviewees indicated that they had some elementary school or, at least, that they knew how to read and write, 23 percent had some secondary or post-elementary education, and, as mentioned before, 11 percent stated that they had never gone to school and did not know how to read or write.

The corresponding figures at the municipal level for 1990 show that 33.5 percent had had some elementary education, 50.1 percent had had some secondary or post-elementary education, while 13.4 percent of the population 15 years old and older had not had any formal education at all.[30]

Backing up the census and the data from my interviews, the salesperson of Vidafel had informed me, during my visit to the resort, that they only employed a "very experienced" labor force. She added that they had hired people from other tourist areas such as Zihuatanejo, Cancún, and Huatulco. Since most of the people in the *ejido* had not had direct experience working in the tourist industry nor did they have the required educational levels,

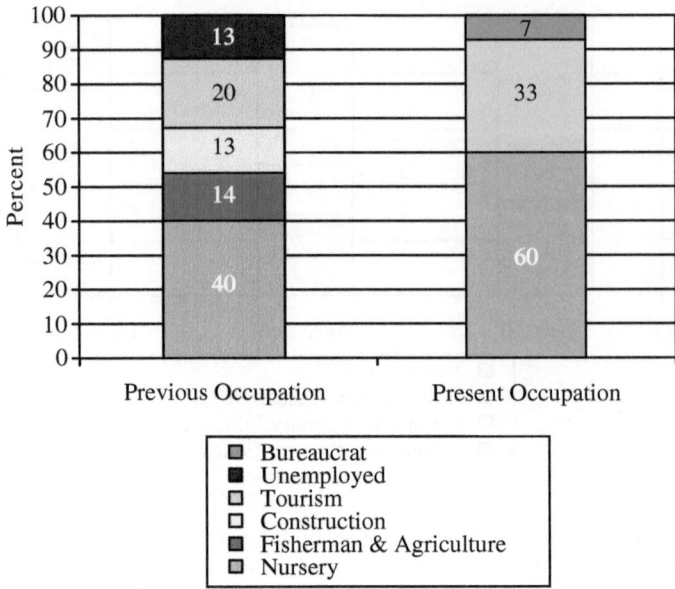

Figure 6.8 Previous and Present Occupations of the Male Labor Force in La Poza, 1994. *Source:* Structured interviews carried out in the *ejido*.

it was evident that *ejidatarios* and their families were not benefiting from the 1,000 jobs allegedly generated by the Vidafel project. Thus the "Happy Life" of Vidafel was not shared by everybody.

THE JOBS WITHIN THE *EJIDO*

The constant expropriations of *ejido* land in La Poza had gradually re-structured their local economy and their income-generating activities. The dominant economic activities in the *ejido* were agriculture and fishery, but *ejidatarios* could no longer sustain those activities as expropriations had lessened the extension and quality of their lands and had prevented their access to the beaches. According to Figure 6.8 occupations of the male labor force in the *ejido* have been very diverse, including a combination of formal and informal activities. About 40 percent had worked in nurseries, 14 percent worked in agriculture or fishing, 13 percent worked in con-struction, 20 percent worked as taxi drivers and other temporary jobs for the tourist industry, and 13 percent were unemployed.

Figure 6.8 indicates that plant nurseries became the survival strategy for the male population in the *ejido*, but the latest expropriation and reloca-tion also reduced the land available for this activity. Nurseries provided ornamental plants for the tourist industry in the area. They also supplied other nurseries outside the state. Among the male population of the *ejido* it was common to observe a variety of employment combinations. Most of them worked on what was left of their nurseries and, at the same time, they were taxi drivers, construction workers, gardeners, painters, and snorkelers. Owing to the decimation of the *ejido*, agriculture and fishing was no longer possible, so that labor force began working in the nurseries. At least 60 percent of the male *ejido* population worked a combination of nursery activity plus another temporary job, while 20 percent worked as taxi drivers and the remaining 20 percent held temporary low-paying jobs in the tourist industry.

Nursery activities in the *ejido* utilized a considerable amount of family labor. Typically, women weeded and watered the plants everyday, while men did the "heavy duty" jobs. When interviewed, most women considered themselves housewives; even though about 55 percent of them participated, in addition to their daily work in nursery-related activities, in some sort of informal income generating activities, such as food and candy preparation. In fact, the percentage of women who just did household work had decreased. When asked about their previous occupations, the female interviewees responded that 28 percent were housewives, 33 percent hold low-paid jobs in the tourist industry, 11 percent hold informal jobs, 11 percent were students, 6 percent worked in the nurseries, and only 11 percent hold formal jobs (see Figure 6.9). In contrast, 17 percent declared that their current occupation was housewife, 22 percent prepared food or candy for sale, and 11 percent work in the nurseries. Currently, female students in the *ejido* accounted for 11 percent, while 6 percent of the women ran their own business, and another 5 percent had a formal wage job. In general, women in this *ejido* had not held any of the few permanent jobs created by the tourism industry.

CERTIFICATES BUT NOT TITLES

Three years after the approval of the 1992 Reforms to Article 27, most of the interviewed *ejidatarios* had no knowledge of those reforms. Just 22 percent had heard about PROCEDE and its implications. I received contradictory statements regarding the implementation of the program in the *ejido*. Some of the *ejidatarios* mentioned that PROCEDE personnel "had measured" the *ejido* parcels. Others indicated that the measuring was taking place only in those areas in which residents had allowed their parcels and lots to be measured. Apparently, some *ejido* members had opposed the program. Nonetheless, the *Comisariado Ejidal* had showed me a letter from CORETT specifying that 38 *ejidatarios* had been granted land certificates. No certificates had been issued for those parcels with limit conflicts. That number of awarded certificates surpassed the original number of *ejidatarios* that the *Comisariado Ejidal* had first given me.

In one of the several interviews with the *Comisariado Ejidal*, he affirmed that a portion of the *ejido* had entered the PROCEDE program in

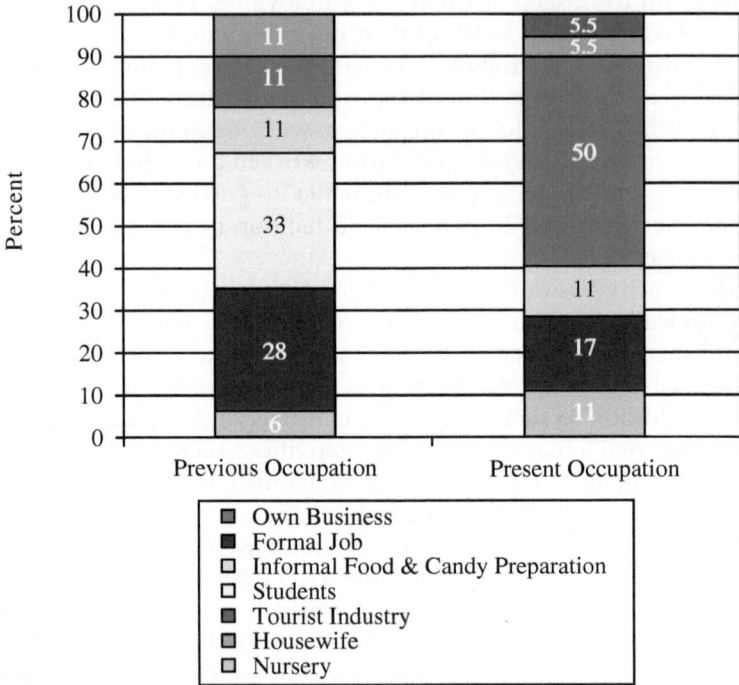

Figure 6.9   Previous and Present Occupations of the Female Labor Force in La Poza, 1994. *Source:* Structured interviews carried out in the *ejido*.

March of 1993. INEGI measured the parceled area for which they awarded certificates. However, residents had refused to have the urban area measured, for which they would have received titles. *Ejidatarios* had correctly considered that the government would start charging property taxes on that land. This was true, *ejido* parcels would continue being un-taxable, but the 1992 *Ejido* Reform allowed the taxation of the *ejido* land devoted to housing. Since Acapulco had "the highest [property taxes] in the country," the *Comisariado Ejidal* said that residents of the *ejido* were afraid to lose their land due to the lack of resources to cover the required taxes. He added that some *ejidatarios* had agreed on participating in the PROCEDE program aiming at the future privatization of the *ejido*. In view of the frequent expropriations of *ejido* land, he considered as urgent the need to privatize it, in order to secure a fair compensation in the future.

The *Comisariado Ejidal* discerned that attaining land certificates and having the option to sell their plots had pros and cons. The positive aspect of attaining land certificates was that *ejidatarios* could sell their unproduc-tive land. According to him, land was unproductive because "people grow little" and "they do not get what they invest." He complained that the *ejido* nurseries were facing strong informal competition from non-unionized nursery workers who sold their plants cheaper than the ones from the *ejido*. Unionized nurseries had started to abandon the business because they were not able to pay back the loans they had contracted with the banks, which were threatening repossession. The negative side of attaining *ejido* land certificates was that the government could expropriate the *ejido* anytime before *ejidatarios* could privatize it and sell it. "I am old, I'd better sell it. I want to die well-fed. Nowadays, money has no value. I would have wanted to leave it [the *ejido*] for them [his sons and daughters], but they study. No matter what, they would end up selling the land. Those who like working it, keep the land. But here [in the *ejido*], we have seen that they [sons and daughters] sell or transfer the land."[31]

The multiple transformations of the *ejido* due, both to expropriations and cultural and generational change had made the *Comisariado Ejidal* very astute. He was witnessing how external and internal forces were transforming the *ejido*; the relocation of capital for tourism purposes and the cultural and generational change within the *ejido* that made young people emigrate or sell the land. His conclusion then was that he better sell the land. That cultural and generational change was also evident in Ixtaltepec, Oaxaca, the third *ejido* studied in this research, an *ejido* that went through PROCEDE without knowing it.

Chapter Seven

Ixtaltepec: An Indigenous *Ejido*

The Ixtaltepec *ejido* had participated in PROCEDE without knowing it. It was not the first time that Ixtaltepec had been involuntarily included in a program. Ixtaltepec—now an *ejido*—used to be an agrarian community, but communal authorities and BANRURAL[1] had worked out an agreement to include Ixtaltepec in an application to change its status from agrarian community to *ejido*.

I learned about Ixtaltepec, Oaxaca during a conference on Indigenous Women and Human Rights that took place in 1994 in Tuxtepec, Oaxaca, a town that borders the state of Veracruz. During the conference, a group of Juchitecan women who had heard about my research invited me to do my fieldwork in an *ejido* of the Isthmus region. They mentioned that some *ejidos* in that area were in the initial steps of PROCEDE, which fit one of my criteria for selecting *ejidos*. Ixtaltepec was an *ejido* close to Juchitán City and belonging to the Isthmus of Tehuantepec region. The cities on both coasts of the Isthmus had received State investment to develop the oil industry, which attracted labor force from the surrounding agrarian communities and *ejidos*. Cultural and generational changes were taking place in Ixtaltepec as younger generations moved out and their absence affected the composition of the population and the ownership of land in Ixtaltepec. Although the *ejido* was not directly receiving the investment for the oil industry, it was indirectly affected by it. Some *ejidatarios* were absentees and some others had started to sell their lands. The 1992 *Ejido* Reforms would hasten this process. Although land privatization took place as in the other *ejidos* of this research, its pace and purpose were different and this is what this chapter explores.

LOCATION AND DESCRIPTION

Ixtaltepec is located in the Southwestern portion of the Tehuantepec Isthmus in the state of Oaxaca, in Southern Mexico. The Ixtaltepec *ejido* is part of the municipality of Asunción Ixtaltepec, which, in turn, belongs to the

143

Juchitán district indicated in Map 7.1. Although Ixtaltepec is only 10 kilometers from Juchitán City, it took a 45 minute bus ride to go from Juchitán
to Ixtaltepec. Road connections, although existing, were in bad shape.

Ixtaltepec is one of nine *ejidos* in the Asunción Ixtaltepec municipality. The municipality of Asunción Ixtaltepec covers approximately

Map 7.1   *Ejido* Ixtaltepec in the Municipality of Juchitán, Oaxaca. *Source:* Instituto
Nacional de Estadística, Geografía e Informática (INEGI), *Juchitán de Zaragoza,
Estado de Oaxaca, Cuaderno Estadístico Municipal, Edición 1995* (Aguascalientes,
México: INEGI, 1995), 9. Based on CGSNEGI Climatological Carte.

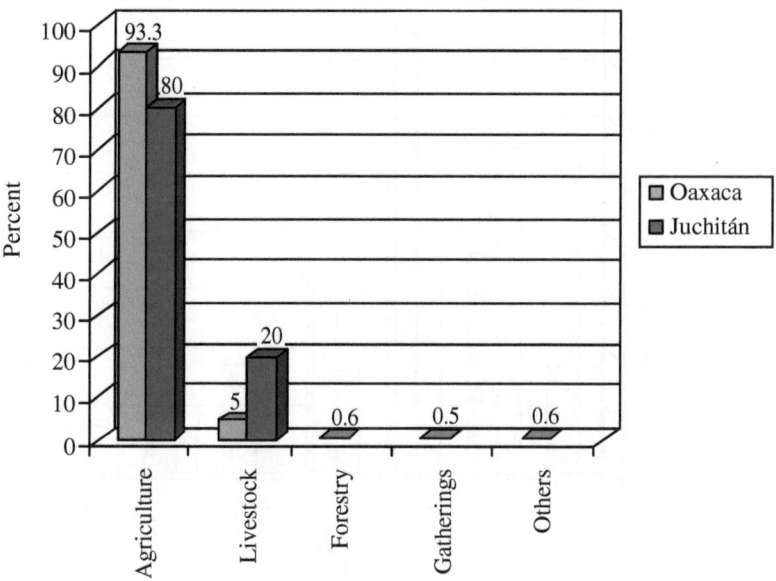

Figure 7.1 *Ejidos* and Agrarian Communities by Main Economic Activity in Oaxaca and Juchitán, 1991. *Source:* Instituto Nacional de Estadística, Geografía e Informática (INEGI), *Juchitán de Zaragoza, Estado de Oaxaca, Cuaderno Estadístico Municipal, Edición 1995* (Aguascalientes, México: INEGI, 1995), 75. Based on data from INEGI, "Oaxaca, Resultados Definitivos. VII Censo Ejidal, 1991."

54,733 square kilometers.[2] The region is one of the windiest in the world, characterized by 45 to 60 kph winds that travel from North to South along the Isthmus region and that occasionally have had adverse effects on the agriculture in the region. Agriculture is considered one of the main economic activities in Ixtaltepec. Among the products grown in Ixtaltepec are corn, sesame, sorghum, peanuts, sugar cane, pumpkin, green peas, watermelon, and cantaloupe. Agriculture is also the economic activity predominant in most of *ejidos* and agrarian communities both in the state of Oaxaca and in the municipality of Juchitán.

As Figure 7.1 shows, Juchitán has a well-developed livestock sector in comparison to the *ejidos* and communities in the rest of Oaxaca. In comparison to the *ejidos* in the San Luis Río Colorado municipality—where 61 percent of them were devoted to economic activities different from agriculture or any other primary sector activity—land in *ejidos* and communities in this region continue to be agriculture-related.

Wage statistics in the municipality of Juchitán indicate for 1990 that 27.2 percent of the population makes less or no minimum wage (18.5 percent receive less than one minimum wage and 8.7 percent receive no wages). Since 37.8 percent of the population receives between one and two minimum wages, this means that 65 percent of the population are low income

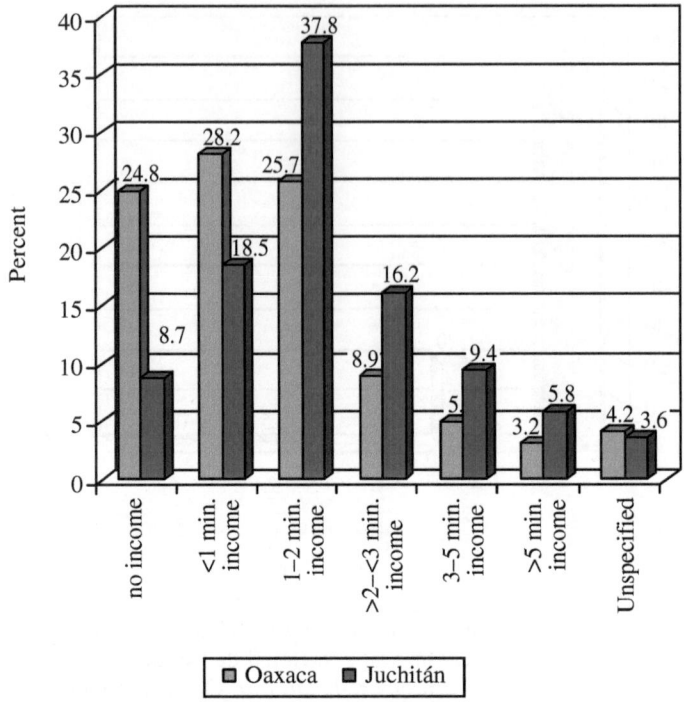

Figure 7.2   Employed Population by Monthly Income in Oaxaca and Juchitán, 1990.
Source: Instituto Nacional de Estadística, Geografía e Informática (INEGI), *Juchitán de Zaragoza, Estado de Oaxaca, Cuaderno Estadístico Municipal, Edición 1995* (Aguascalientes, México: INEGI, 1995), 66. Based on data from INEGI, "Oaxaca, Resultados Definitivos. XI Censo General de Población y Vivienda, 1990."

(see Figure 7.2), in comparison to the 43.3 percent low income population in SLRC.

In Ixtaltepec, just 35 percent of its population receives 3 times the minimum wage of more while 56.7 percent of the population of SLRC, in the North, received those wages.

Although the main economic activities in *ejidos* and agrarian communities are agriculture and livestock, the census data shows that Juchitán is predominantly urban. According to the Juchitán Municipal Statistical Book, the municipality has a predominately urban population (see Figure 7.3).

This urban predominance contradicts its data on main economic activities as well as my observations in the *ejido*, which I found was mostly rural and obviously does not correspond with the composition of the rural and urban population of the municipality of Juchitán to which Ixtaltepec belongs. However, as warned in Chapter Five, this data should be considered with reservations as the definition of urban includes those localities with more than 2,500 inhabitants regardless of other characteristics.

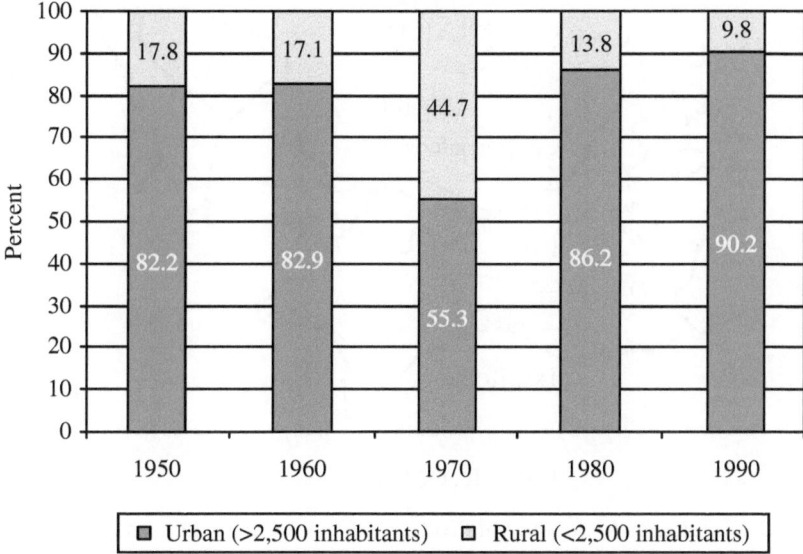

Figure 7.3 Urban and Rural Population in Juchitán, 1950–1990. *Source:* Instituto Nacional de Estadística, Geografía e Informática (INEGI), *Juchitán de Zaragoza, Estado de Oaxaca, Cuaderno Estadístico Municipal, Edición 1995* (Aguascalientes, México: INEGI, 1995), 16. Based on data from INEGI, "Oaxaca, Resultados Definitivos. VII–XI Censos Generales de Población y Vivienda, 1950–1990."

The Isthmus region has been influenced over time by different investment ventures. First, the Transisthmian Railroad, built at the beginning of this century, served as the connection between the Atlantic and the Pacific coasts. With the construction of the Panama Canal, the area's economy declined.[3] In the 1940s, the oil industry in the region attracted labor from nearby towns and cities. The population from Ixtaltepec also migrated to oil ports such as Salina Cruz on the coast of Oaxaca and Coatzacoalcos on the Atlantic coast (see Map 7.2).[4]

In contrast to SLRC which enjoyed most urban services, in this *ejido* conditions were precarious. Although they had electricity, they did not have drinking water and, in many cases, they did not use gas for cooking. Approximately 44 percent of the women told me that they used firewood that their family collected. They drank water from the wells each family had in their yard. The water was pumped through hoses that were connected to faucets. Since water had been the main cause of infectious gastrointestinal and skin diseases in the area, they sometimes boiled the water or added chlorine tablets in order to improve the water quality. Those families with greater resources— about 22 percent of my interviewees—told me they bought drinking water.

Indigenous communities have had a stronger presence in Southern Mexico than in the rest of the Mexican Territory. Oaxaca has a strong

Map 7.2   Ixtaltepec *Ejido* and Oil Producing Cities. *Source:* Drawn by author.

sense of pride in its Indigenous cultures, languages, and ways of living. Most of the Isthmus population is of *Zapoteco* origin and the majority speaks *Zapoteco*. In 1990, approximately 76 percent of the population above 5 years old spoke an Indigenous language in the municipality of Juchitán alone.Of the Indigenous speaking population, 97 percent spoke *Zapoteco*.[5] I assume that the proportion of Zapotecan speakers in Asunción Ixtaltepec, and in the Ixtaltepec *ejido*, is about the same or higher as about 100 percent of my interviewees greeted me in *Zapoteco* when I first met them and 85 percent reported that they speak *Zapoteco* at home.[6]

The population data I was able to gather for the *ejido* was extremely contradictory. For example, an anonymous report provided by the *ejido* authority specified the total population of the Asunción Ixtaltepec municipality as being 32,154 inhabitants. On the other hand, a statistical data sheet also provided by the *ejido* authority indicated that the population of the municipality was 3,039 inhabitants. However, neither the report, nor the statistical sheet gave the year of the data or its source. It was difficult to estimate the population of the *ejido*, since the list of *ejidatarios* that the *Comisariado Ejidal* had was outdated. Although his roster listed 636 *ejidatarios*, most of them were not living in the *ejido*. Apparently there was evidence of absentee *ejidatarios*.

HOW IXTALTEPEC BECAME AN *EJIDO*

Ejidos in Oaxaca are not common. Most of its land, about 5,399,883 hectares, has historically been organized as *tierras comunales* (communal lands) as they called them. As explained in Chapter Two, *tierras comunales* have been recognized as belonging to Indigenous communities since the time of the Spanish Crown. The state of Oaxaca has the highest number of agrarian communities and comuneros in Mexico. In 1991, it had 1,418 agrarian communities, of which 684 were *tierras comunales*, representing 48 percent of the total agrarian communities in the state.[7] About 190,000 comuneros owned those lands. Asunción Ixtaltepec stands alone, having had its communal lands converted into *ejidos* in 1963. It is important to note, however, that the *comuneros*, now *ejidatarios*, did not realize that fact for a long time, since it was not they who had requested the conversion.

An official monographic study of 1975 provided by a retired teacher in the *ejido*, indicates that in 1954 during the Ruiz Cortínez administration, the land rights of the residents of Asunción Ixtaltepec had been confirmed and titled as *tierras comunales*. This legal decision was enacted in 1956.[8] However, by the time that monograph was written, many of the *ejidatarios* still did not know that somehow, somebody had changed the status of their *tierras comunales* to that of *ejido*. The monograph indicates that the change in land tenure was announced in the *Diario Oficial de la Nación* (the National Official Newspaper) on November 30, 1963, during the López Mateos presidency. The *Diario* announcement emphasized that the land tenure change had been carried out at the request of *comuneros*. However, *comuneros* insisted that they did not know of, nor had they requested, such a change.

Evidently, the change was made by the authorities of the community without consulting all the *comuneros*. Since the then recently built Benito Juárez dam was going to provide irrigation benefits to the lands surrounding the area, it is possible to infer that some kind of hidden agreement between communal authorities and public officials of the *Banco Nacional de Crédito Ejidal* (BANRURAL) or National Bank of *Ejidal* Credit took place. The basis for this reasoning is the fact that the communal lands, once incorporated as *ejidos* within Irrigation District 19—the irrigation jurisdiction of the Benito Júarez dam—were eligible to receive loans from the Bank. This credit was intended to support the growing of export commercial crops. Another element that supports this reasoning is the fact that BANRURAL initiated an association with the *comuneros* (many of whom still did not know that they had been converted to *ejidatarios*) in order to encourage them to start growing rice for export. As part of that agreement, BANRURAL requested that *comuneros* elect a representative who would then be the only person from the community with whom BANRURAL would deal about credit, rice production, land, and profits.

This was a very questionable agreement since *comuneros/ejidatarios* had to turn over their land to BANRURAL to grow rice. BANRURAL provided the credit, but also dictated how it was to be used. In addition, *comuneros/ejidatarios* were not allowed to work on their lands at all and, at the end of the cropping season, it was BANRURAL alone which decided the amount of profit to be distributed to the peasants, who, in many cases, did not receive anything. *Comuneros* never knew how much credit was received, how it was allocated, how much rice was harvested, or where it was sold. Some of them complained that they never even learned the techniques for growing rice.[9] *Técnicos*, as peasants called the agrarian technicians, "were in charge of doing everything." Basically, what BANRURAL had done was to request the credit in the name of the *ejido* and then invest the credit in an export-commercial crop, in this case, rice. BANRURAL, in conjunction with communal authorities, was exploiting the *ejidatarios* who received minimal or no profits. Because of the irregularity of the whole process, some of the *ejidatarios* demanded a report of credit and production activities and of the benefits received by each and all of the participants in the association. Evidently, BANRURAL did not provide any details about the enterprise and, in protest, some of the *ejidatarios* withdrew their lands from the association. BANRURAL responded by stopping credit for those peasants. As peasants had not received any report about the activities of BANRURAL, they appealed to a higher authority in the form of the Ministry of Agrarian Reform, but received no answer. Since then, many of the peasants see credit as one of the ways to take advantage of them, and some of them refuse to see themselves as *ejidatarios*, but rather as *comuneros*, as *ejido* for them, had taken on too many bad connotations. In contrast to the SLRC *ejido* that successfully halted its relation with BANRURAL,[10] the *ejidatarios* in Ixtaltepec were severely affected by their protest against BANRURAL as they did not have the same access to financial resources and independence in decision-making as did the Northern *ejido* of SLRC.

THE *COMISARIADO EJIDAL*

When I first arrived at the *ejido*, I went to the house of the *Comisariado Ejidal* to introduce myself and let him know that I was going to be working in the *ejido*, undertaking research. I first introduced myself to his wife Francisca, who told me that I should wait for the *Comisariado* because he was in a meeting. When the *Comisariado* arrived, he immediately asked me if I wanted to buy some land in the *ejido*. It was my first awareness of the fact that land sales were taking place in the *ejido*.

I let him know that I was undertaking research on *ejidos* with the objective of finding out what impact, if any, the 1992 Article 27 was having. I also let him know that I intended to conduct interviews with families in the *ejido*. At this point he assigned his niece, who was also his administrative assistant for the business related to the *ejido*, to accompany me in all my interviews and visits in the *ejido*. At first I saw her presence as a restriction to my communication with *ejidatarios* and their families. In fact, she was of invaluable help, as she became my interpreter when I came across people whose only language was *Zapoteco* or who preferred to speak *Zapoteco* over Spanish—some 85 percent of my interviewees.

PROPERTY IN IXTALTEPEC

During an interview with the then *Comisariado Ejidal*, Hiram Martínez Antonio stated that there were private, communal, and *ejido* lands coexisting in Ixtaltepec. However, the 1980 Census had reported no private or communal lands in the municipality. This contradictory information suggested to me that some *ejidatarios* had declared themselves as *comuneros* because, either they did not know their legal status or did not consider themselves as *ejidatarios*, although legally the conversion had taken place. That different forms of property actually coexisted also pointed to the sale of *ejido*/communal lands. According to the 1975 monograph, peasants had declared that they owned *ejido* land as well as private land and that they did not pay any taxes on either form of property. Since, by law, *ejidos* are property-tax exempt, it is my guess that informal sales of *ejido* land took place between and among *ejidatarios* (and possibly non *ejidatarios*, as well) and, although they considered that property as privately owned, because it had not been awarded by the State, but had been purchased, legally it continued to be under the jurisdiction of the *ejido* system.

LAND, LABOR, AND MIGRATION

The 1991 Ixtaltepec *ejido* roster listed 636 *ejidatarios*, of whom twelve were women. The roster did not indicate the amount of land *ejidatarios* owned or to which of the nine *ejidos* that the Census listed they belonged. According to the monograph provided by the retired teacher, the extension of most of the parcels ranged from three to five hectares, while a few privileged parcels were as large as 20 to 100 hectares. Neither the monograph nor the *Comisariado Ejidal* could tell me how some people had accumulated so much land within a regime that was supposedly designed to avoid land concentration.

According to the monograph, during the rice-credit scam, the bigger the landowner the more credit they received. In contrast, the majority of peasants whose parcels were small were further exploited by that scam. The insufficient size of their parcel, along with periods of strong winds and long droughts, had led some of the peasants to look for other income-generating activities. In addition to agriculture, the population of the *ejido* began working in the pottery industry. This was an almost exclusively male activity.

In reviewing the roster of *ejidatarios* to help me select a sample of *ejido* families to interview, the *Comisariado Ejidal* warned me that it was not going to be possible for me to meet most of the people on the list. Without specifying how many, he explained that a number of them had sold their *ejidos*, some had migrated out of Ixtaltepec, and some others had died and their descendants no longer lived in Ixtaltepec. Of the first eight *ejidatarios*, chosen randomly from the list, only one still lived in the *ejido*. The information on land and population that appeared in the roster differed greatly from the actual population living in the *ejido*. Changes in population and land had taken place over time. In and out migration and land privatization had been occurring even before the implementation of the new law. Due to the unreliability of the *ejido* records, I decided to approach my interviews without the use of the roster, going house by house. This strategy meant I might be interviewing everybody instead of just *ejidatarios/comuneros*.

According to the *Comisariado Ejidal*, those people who had sold their *ejidos* migrated to Mexico City, Salina Cruz, or Veracruz. Both Salina Cruz and the coast of Veracruz belong to the oil producing areas. Both areas were growing rapidly during the 1970s to1980s and provided an economic alternative for labor in Ixtaltepec. Using the roster as a basis to determine the percentage of population that had moved out along with their families, it seems that about 70 percent had moved to other regions. The *Comisariado* mentioned that in times of economic crisis, the oil producing regions had fired many of those ex-*ejidatarios/comuneros*. Some of them returned to Ixtaltepec, but since they had sold their lands, they no longer had resources. Some others stayed in Salina Cruz or Veracruz because they saw no point in coming back to Ixtaltepec as landless people.

A generational element also played a role in the transformation of the *ejido*. Youth who had had the resources to leave, either in search of employment or educational opportunities, left behind their parents or relatives. They were usually not interested in coming back to Ixtaltepec to work as farmers. Culturally, they were no longer *ejidatarios* and their idea of well-being did not include living in a rural area with no basic services, at the mercy of corrupt authorities and Agrarian bureaucrats. Interestingly, the Ixtaltepec monograph indicates that young women in Ixtaltepec did not leave as readily as did the young men.

THE LOCAL ECONOMY

During my staying in Ixtaltepec, I lived at the home of the *Comisariado Ejidal* and his wife Francisca where I soon became aware of the way women support the local economy. Francisca would get up every morning between and 3:30 and 4:00 in order to go to the *molino de nixtamal*,[11] a corn-grinding mill. Upon returning home, she prepared *atole*[12] with that *nixtamal*.[13] By 7:00 or 7:15 A.M. several women and children would arrive to buy *atole* from her. These would bring to sell or exchange other products such as dried shrimp, smoked and fresh fish, homemade bread and cookies, chocolate, *quesadillas*, *tortillas*, meat, chicken, and other produce and prepared food. At the same time, other women would offer their services as housecleaners and launderwomen. Each day, the home of the *Comisariado Ejidal* became the gathering point for those women who started their routine by buying *atole* and selling and exchanging their products and services. This aspect of the *ejido*'s local economy displayed, on the one hand, a partial gender aspect of the labor force in the *ejido*, and, on the other hand, it showed the nonmonetized or semi-monetized economy in which the *ejido* female population engaged while their husbands worked in agricultural or pottery activities.

CULTURAL TRANSFORMATION IN THE *EJIDO*

Three days before starting my stay at Ixtaltepec, I was invited to a traditional wedding in Juchitán City. By witnessing that wedding, I understood how couples and families survived as a result of a strong moral economy that still exists in the area. Most guests brought food and gifts and gave money to the couple. Nonetheless, people in the party complained about how weddings today were different from in the past. They noted that the couple used a plastic cover instead of palm leaves to build the tent for the party because it was more affordable; they did not hire a marimba ensemble to play, but hired instead a regular city music group, which was cheaper; and the bride was not wearing the traditional *Tehuana*[14] wedding dress because it was more expensive than buying a regular Western wedding dress. It was simply not affordable to keep some of their traditions anymore, although they tried to keep as many as possible, maintaining those traditions today meant that they were forced to substitute clothing, music, and other ancillary materials which at least gave a semblance of old tradition, in spite of their change in form.

The wedding was also an illustration of the changing nature of the regional culture and its population. The older women were wearing their traditional dresses, the younger women were not. I was told that many younger women, with access to education, leave for other Oaxacan cities because in Juchitán and its surrounding cities there are no universities. The Isthmus population felt that this change of residence had a profound effect on the way women dress and live.

During the wedding in Juchitán, men sat on one edge of the patio and women sat on the opposite edge, in front of the men. That was the customary gender arrangement in gatherings and parties. While men talked and drank among themselves, women danced with each other. Rogelia, my Juchitecan friend, told me they were "fed up" with all the articles in Mexico City's magazines that portrayed women from the Isthmus as "feminists" and economically supporting their men. She mentioned that women writers from those magazines would come stay a couple of days, observe a few people, and draw conclusions without asking about the motives for certain behaviors.

A reporter from *Ella* magazine, for example, wrote that Isthmus women worked hard while men spent their time drinking and sleeping.[15] Rogelia told me that several women from Juchitán had sent a letter to the magazine to protest the distorted image those articles portrayed. She said that men could not work all the year round because of the nature of the local economies. Most men work in agriculture or activities related to agriculture, which is mainly seasonal work. In addition, the regional rates of unemployment for men in both non-agricultural and agricultural activities were high. For women, the situation was different; they were able to work as traders or merchants throughout the year. That was not a reason, she argued, to conclude that men did not work at all and that women supported them; rather, it was a question of the nature and seasonality of those activities and, I would add, of the gender division of labor in the *ejido*.

PROFILE OF THE POPULATION IN THE *EJIDO*

In Ixtaltepec, I interviewed families in the *ejido* as well as some elderly people. At the time of my visit, most women were by themselves, taking care of their homes and their children, and they informed me that their partners were working in the fields. Women did not work in agriculture. The results of my questionnaire indicated that the age of those women ranged from 40 to 63 years, which indicated that a considerable migration of women between the ages of 17-40 years old had taken place.

Most of my interviewees and their partners had been born in Ixtaltepec as had their parents and grandparents. About 20 percent of them had been born in small towns outside the *ejido*, but within the Ixtaltepec municipality. The parents of this immigrant population had also been born in other small towns within the Isthmus region, which indicates that in-migration for the population in Ixtaltepec was not a significant trend or, in cases where it occurred, it had been mostly intra-migration.

Out-migration, however, was considerable. Although residents in the 40-63 year old group rarely emigrated, those between 18-25 years old regularly left for Oaxaca City to pursue higher education, or to other cities, such as Juchitán, Salina Cruz, Tuxtla Gutiérrez, Matías Romero, Veracruz, and Mexico City, in search of better-paying jobs. It was evident that the working population currently living in the *ejido* was mainly composed of men and women 40 years old or older. The results of my interviews had confirmed what the *Comisariado Ejidal* had told me. Younger generations migrate out and do not return to Ixtaltepec after completing their education or work, and the few who return, do not work in agriculture.

About 73 percent of women interviewed were married and self-employed. Only 13 percent were heads of households. The family structure in this community was more cohesive and supportive than in the other two *ejidos* of this research. The average number of children born to a woman was 4.3. Those women with no partners had informally adopted children from the community. Households were composed of about 5.6 members. This figure reflected a variety of family arrangements. For example, in some households, children had migrated out or married, so they no longer lived in the same household as their parents. In the cases when children stayed, they were integrated into their parents' household or into other relatives' households in the same *ejido*. Some households were composed of a mix of families. Two notable cases were that of a divorced, childless woman who had adopted 5 sons and daughters and 2 nephews who had lost their parents, and whose father lived in the same house, and another of a single mother who had adopted an additional child; her household was composed of 9 other members of her extended family, including dad, her sister's family, and a cousin. These women had more resources and education than the rest of my interviewees. One of them was a school teacher and the other one had studied until 5th grade and owned a small business.

About 40 percent of the population I interviewed had not attended school beyond 3rd grade. I was told by one of the elementary teachers that elementary school was only available through third grade, so it was not possible for *ejido* children to study beyond that grade. As a result, the educational level of most of the residents in Ixtaltepec reflected that limitation (unless there were people with resources that could migrate to other cities for education.) Another 40 percent had no schooling at all and did not know how to read and write. This percentage was high in comparison to those of the state and the municipality with 26 and 24.7 percent, respectively (see Figure 7.4).

Generally, the inhabitants with highest education were the school teachers and the *ejido* assistant, who had clerical and accounting training. These numbers indicate that both the state and the municipality had higher

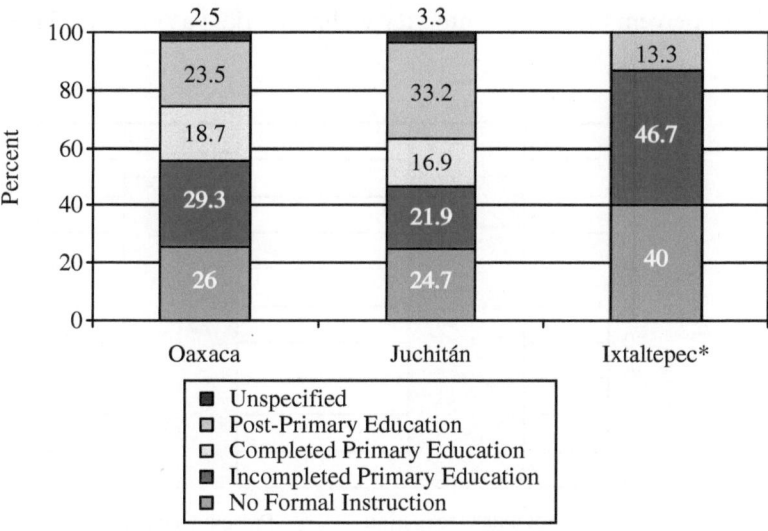

Figure 7.4 Population 15 Years and Older by Educational Level in Oaxaca, Juchitán, and Ixtaltepec, 1990. *Data obtained from the results of the structured interviews carried out in the *ejido* by author in 1994–1995. *Source:* Instituto Nacional de Estadística, Geografía e Informática (INEGI), *Juchitán de Zaragoza, Estado de Oaxaca, Cuaderno Estadístico Municipal, Edición 1995* (Aguascalientes, México: INEGI, 1995), 46. Based on data from INEGI, "Oaxaca, Resultados Definitivos. XI Censo General de Población y Vivienda, 1990."

educational levels than those in the Ixtaltepec *ejido*. In Oaxaca and Juchitán, 74 percent and 75.3 percent respectively, had received some kind of
education, while in the *ejido* just 60 percent of the population had received
education. The number of people with no instruction in the *ejido* was
higher than at the state and municipal levels.

LOCAL LABOR MARKETS

In order to historically trace the work that the Ixtaltepec population has
engaged in over the years, residents were asked about their previous and
current occupations inside and outside the *ejido*. They were also asked
about the occupations of their parents and grandparents. The information
obtained revealed that most of the economic activities still being carried
out by both women and men in the *ejido* had been inherited from their parents or other relatives. However, some restructuring in the labor markets in
the region had affected the occupations in the *ejido*. The fairly young oil
industry in Salina Cruz, Minatitlán, and Coatzacoalcos had attracted male
labor from the entire Isthmus region, including Ixtaltepec. Other factors
contributing to the restructuring of the local labor markets in the *ejido*
were detrimental agricultural policies, the degradation of the environment,
and climate changes that had mostly affected agricultural work.

Figure 7.5 shows previous and current occupations of men in the *ejido*.
While 54 percent of the men interviewed had worked as farmers, 31 percent

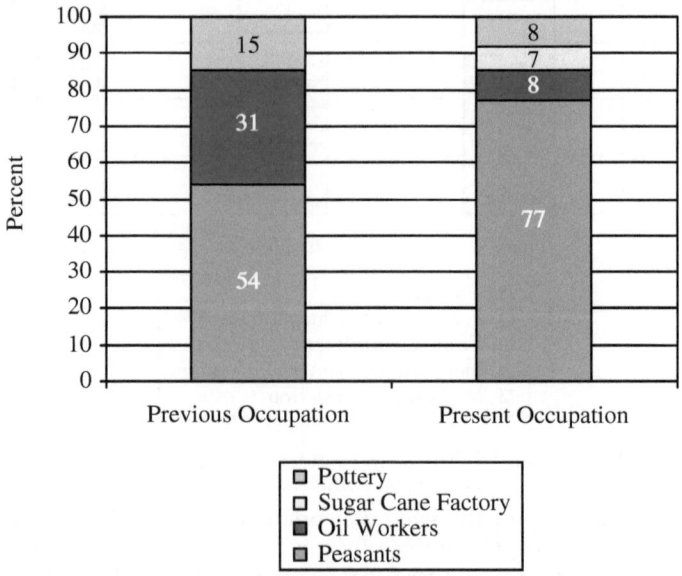

Figure 7.5   Previous and Present Occupations of the Male Labor Force in Ixtaltepec,
1994. *Source:* Structured interviews carried out in the *ejido*.

had been employed as oil workers outside Ixtaltepec, and 15 percent had worked making pottery. When they were asked about their current occupations, they stated that 77 percent worked as farmers, 8 percent worked doing pottery, another 8 percent still worked in the oil industry, and 7 percent worked in a sugar cane factory that was in the process of being closed because of privatization of the sugar cane industry affecting this sector.

These figures indicate that the oil-producing cities attracted people from Ixtaltepec, but apparently did not retain all of them. Since most of the working population in the *ejido* is over 40, it is possible to conclude that a younger labor force displaced older labor in the oil industry and the older workers then returned to the *ejido* to work again as farmers.

Residents in the *ejido* reported that 67 percent of their fathers worked as farmers, 20 percent worked in the pottery industry, and 13 percent worked in non agricultural jobs, which are related to the economic and cultural life of the *ejido*. In general, the grandfathers of the interviewees worked in agriculture or by making pottery, basically the same employment as their sons continue in today.

Although male labor markets in the *ejido* have changed slightly, no major economic changes in the *ejido* female labor markets were reported. When I first interviewed women about their occupations, about 90 percent of them responded that they were housewives. It was not until I asked them about their daily routine that they mentioned other economic activities. Initially, they had not reported these activities as income-generating because they did not consider them as work. The common thread linking all those activities is trade. Their grandmothers had been traders, their mothers were traders, and they too were traders. The main product they traded was food, both crude and prepared. Reyna Aoyama reports that women traders developed this skill when economic crisis pushed them to economically contribute to support their households.[16] She also explained that the Trans-isthmian railroad promoted this activity.[17] However, in our interviews, women told me that trading and food preparation were activities undertaken in the *ejido* even before the completion in 1960 of the railroad that connected the oil cities of Coatzacoalcos and Minatitlán in Veracruz, and Salina Cruz in Oaxaca.[18]

As shown in Figure 7.6, historically the predominant female economic activity was trade; 47 percent of the female labor force in the *ejido* was accustomed to doing some kind of food preparation for sale or trade. Another 26 percent reported they were exclusively housewives, while 13 percent were employed in the pottery industry. 7 percent were farmers and, the remaining 7 percent stated that they worked in some service industry.

When asked about their current employment, 47 percent of women continue to work as traders of food and produce, 20 percent now worked in the service industry, 13 percent were housewives, and another 13 percent worked as farmers. In addition, 7 percent of them worked in the pottery industry. These figures indicate increases of 13 percent and 7 percent in the female labor force working in the service industry, and in agriculture,

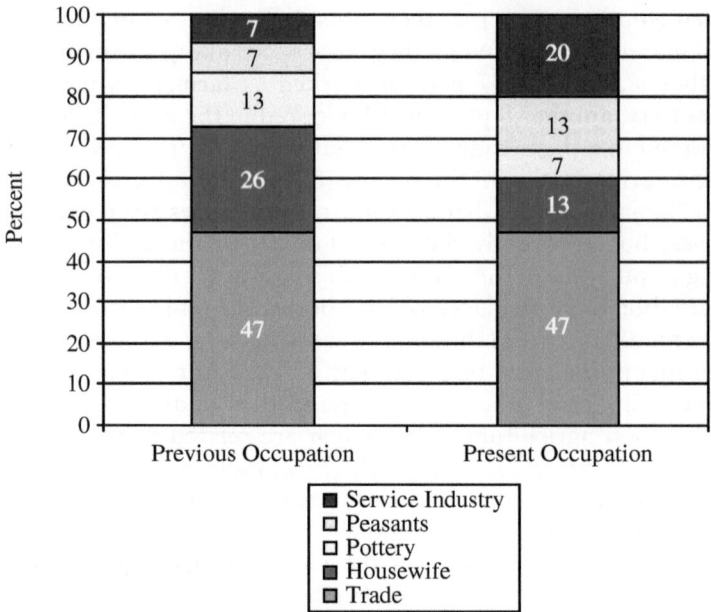

Figure 7.6 Previous and Present Occupations of the Female Labor Force in Ixtaltepec, 1994. *Source:* Structured interviews carried out in the *ejido*.

respectively. They also display a 6 percent decrease of women devoted to pottery activities.

Women in the *ejido* reported that 73 percent of their mothers had worked as traders of food and produce, while 20 percent had been housewives and 7 percent had worked in the pottery industry. The only reported change was that, today, 13 percent of the women traders do embroidery, instead of trade. Thus, although they do not trade food and produce, they still trade their embroidery work. Women reported no variation in the activities of those women working at home as housewives or in the pottery industry. These employment trends in Ixtaltepec showed a different direction from the municipal and state employment figures of 1990, which even among themselves showed discrepancies.

According to Figure 7.7, at the state level, the primary sector was 52.9 percent of the total economy of Oaxaca, while the secondary sector accounted for 16.4 percent and the tertiary sector amounted to 28.3 percent. At the municipal level, however, the primary sector accounted for only 23.7 percent of the total economy of the Municipality of Juchitán, while the secondary sector was 30.9 percent and the tertiary sector was 43 percent. Although slightly dissimilar, the state economy is more like the local economy in Ixtaltepec than the municipal economy of Juchitán. This means that the state of Oaxaca continues being a predominantly agricultural state as does

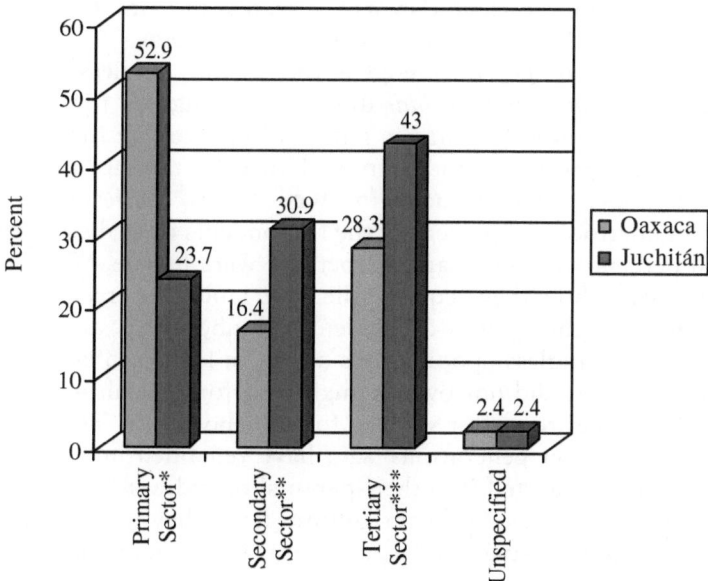

Figure 7.7 Employed Population by Economic Sector in Oaxaca and Juchitán, 1990. *Includes agriculture, livestock, forestry, and fishing. **Includes mining, oil and gas extraction, manufacturing industry, generation of electric energy, and construction. ***Includes commerce and services. *Source:* Instituto Nacional de Estadística, Geografía e Informática (INEGI), *Juchitán de Zaragoza, Estado de Oaxaca, Cuaderno Estadístico Municipal, Edición 1995* (Aguascalientes, México: INEGI, 1995), 65. Based on data from INEGI, "Oaxaca, Resultados Definitivos. XI Censo General de Población y Vivienda, 1990."

the economy of Ixtaltepec. The dissimilarity of the figures for the municipality of Juchitán is explained by the rapid transformation of Juchitán City and the lands around it into urbanized lands.

EJIDO LAND IN IXTALTEPEC

It was not possible to get exact records either on the number of people who had sold their land or those who had held on to it as the *Comisariado Ejidal* did not have such records. The only piece of information that he had available was an outdated roster of *ejidatarios*. Nonetheless, in my interview with him, he did indicate that some *ejidatarios* had sold their land, migrated out, and never returned. Others, he claimed, despite having sold their land, returned to Ixtaltepec. This latter group, once in Ixtaltepec, either bought, rented, borrowed land, or worked for somebody else. On the other hand, some of the migrants who had left Ixtaltepec, did not sell their land. Thus, when they became unemployed either in the city or in the oil-producing region to which they had migrated, they still had the option of returning to Ixtaltepec to resume working as farmers.

Another group to be considered is that of the landless population. Although the *Comisariado Ejidal* did not have figures for this group, from my visits to several *ejidatario* families I believe that the majority of people in Ixtaltepec were land owners. However, the fact remains that there are landless people in the *ejido*. According to the results of the interviews I undertook, 73 percent of the respondents owned *ejido* land. Of these, 60 percent were male and 13 percent were female. When respondents were asked how they acquired the *ejido* land, 53 percent responded that they had bought it, while 20 percent had inherited it, and 27 percent were landless. "Landless" populations might, in fact, own land on which they lived, but they did not own enough land for agricultural purposes. To farm they either rented or worked for somebody else. This is particularly true of younger generations who have remained in the *ejido*, but who have been integrated into their parents' or in laws' households. Although, they do not own *ejido* agricultural land, they were often able to purchase land for housing from a relative or from people moving out of Ixtaltepec.

In summary, *ejido* land sales allowed new generations to have access to land for housing as the *ejido* subdivided, or they acquired it from those who migrated out. In fact, 33 percent of the respondents who owned *ejido* land indicated that they had acquired it from relatives. A comment by one of the respondents is very telling about this situation: "It is not *ejido*. We bought it from my Dad. It is private property."[19] In this context, it is important to emphasize that based on the first search of current *ejidatarios* residing in the *ejido*, only 13 percent of the total *ejidatarios* listed in the 1991 roster, still lived in the *ejido*. Another element to be highlighted is the fact that privatization was taking place even before Article 27. Therefore, although before 1992, it was not legally possible to alienate *ejido* land, purchases took place and, in the minds of the purchasers those were not *ejido* lands anymore, but private property as a price had been paid for them.

As regards land ownership it is also interesting to note that 47 percent of the female respondents to my interview considered that they owned land along with their husbands, even if men were the "official" owners of the *ejido*. Although they did not work the land, they used its products in order to carry out their trading activities.

LOCAL LABOR FORCE AFFECTED BY CHANGES IN *EJIDO* LAND TENURE

Although the 1992 Article 27 posed a potential loss of agricultural land as it allowed a new wave of privatization of *ejido* land, its impact was not immediately clear. *Ejidatarios* continued working as farmers and women continued trading and traveling to sell food and produce. Apparently, the local labor force attached to the economic base in the *ejido*, was not at risk of displacement. It did, however, face other risks, such as lack of credit, dishonest state authorities, and a growing concentration of *ejido* land in few hands. In Ixtaltepec, the forces most affecting the character and development of the *ejido* were internal. It is true that the Coatzacoalcos-Minatitlán and the Salina Cruz oil producing regions attracted labor from Ixtaltepec, but in part, those migrant populations also left Ixtaltepec because of the situation within the *ejido* that contributed to making it more profitable and convenient for some *ejidatarios* to move out.

Changes in *ejido* land tenure were not accelerated or visible. There was some sale of *ejido* land but this was occurring mostly inside the *ejido* and few transactions took place outside the *ejido*. The process, thus, was one of some landlords accumulating more land, and some *ejidatarios* selling their land before leaving the *ejido*, moving permanently to other labor markets and other living spaces.

DO YOU KNOW ABOUT PROCEDE?

None of the families I interviewed knew about the 1992 reforms to Article 27 and the PROCEDE program implemented by the *Procuraduria Agraria*. However, they informed me that the INEGI personnel had started measuring their parcels. They believed that the measuring was for their benefit as it would help them be sure what land belonged to them. They did not know, however, that the measuring process was the first step in including the *ejido* in the PROCEDE program and that, in order to be included, *ejidatarios* by law must have agreed to and voted on it during an *ejido* meeting. They were never informed that they had the option of not participating in the PROCEDE program. I asked the *Comisariado Ejidal, ejidatarios* and personnel of the *Procuraduría Agraria*, about the application of PROCEDE in Ixtaltepec, but I could not find out how INEGI had started the measuring process without having followed the proper procedures.

As I was leaving the *ejido*, I learned that newly-elected *ejido* authorities had authorized the PROCEDE program without informing everybody of its possible consequences. Japanese capital wanted to take advantage of the winds in the region to produce energy by installing wind mills. An employee of the *Procuraduría Agraria* told me that, in order to accomplish this investment project, the *ejido* needed to privatize its land. However, not all *ejidatarios* had been informed about the project or the land certification program itself. Lack of information and lack of required consensus made PROCEDE appear as another land policy instrumented in an authoritarian fashion. Although PROCEDE was not the force leading *ejidatarios* to privatize their lands, it legitimated a process already occurring at different degrees in each different region. The next chapter concludes this research by comparing the three *ejidos* studied in Chapters Five through Seven and contrasting their privatizations and the meaning of those privatizations for planning.

Chapter Eight

Comparing the Three
Ejidos/Comparing Privatizations

The three *ejidos* studied in Chapters Five through Seven of this research—San Luis Río Colorado, La Poza, and Ixtaltepec—indicate the current status of land policies and regional trajectories in Mexico as well as the different actors related to the land of these *ejidos*. As stated in Chapter One, the element that prompted this study was the approval in 1992 of the *Ejido* Reforms, also called 1992 Article 27. This legislation motivated me to study *ejidos* to find out its impact by looking at their internal and external relations and their relation to privatization, urbanization, economic restructuring, and the creation of new investment spaces. This perspective took me to the history of land and *ejidos* in Mexico, the social relations established in *ejidos*, and the importance of their locations, which implied going beyond the sphere of agricultural land and into the arena of urban land and economic spaces.

This final chapter compares the three *ejidos* and returns to the initial questions that directed this research. Is the new legislation fostering further privatization of *ejido* land? Is *ejido* land related to urbanization, the formation of regions, and the currently so-called globalization? This chapter reviews the findings to answer those questions which are related to land conversion and privatization, the labor force transformations that land conversion brought about and, most important, the rural-urban transformation that this policy has prompted by allowing national and transnational capital investment to change the landscape of Mexican *ejidos*. In order to frame the content of this chapter, Table 8.1 presents again a summary of the major findings in each *ejido* arranged first by the selection criteria, that is, by region, geography, regional economic base, and their participation in PROCEDE. Then it describes whether land privatization and land use changes were taking place within the *ejido*. It reviews transformation in labor for both the male and female population and, finally, it describes the impacts on land use.

TABLE 8.1 Three *Ejidos* in Three Different Regions (Recapitulation)

	San Luis Río Colorado, Sonora	La Poza, Guerrero	Ixtaltepec, Oaxaca
Region	Northwest	Centralwest	Southwest
Geography	Border	Coastal	Isthmian
Regional Economic Base	Maquiladora	Tourism	Oil, Agriculture
PROCEDE	Privatization	Refused to participate	Without consent
Land Privatization	Illegal sale for housing Expropriation for housing and manufacturing	Illegal sale for housing Expropriation for tourism	Illegal sale for housing
Labor Before			
1. Male	Agriculture	Agriculture & fishing	Agriculture & pottery
2. Female	Agriculture	Plant nurseries	Informal trade, food
Labor After			
1. Male	Agriculture & services	Services, tourism industry	Agriculture & oil industry
2. Female	Maquiladora industry	House cleaners, maids	Informal trade, pottery
Impact on Land	Land for industrial park. Conversion for housing and services	Land for tourist resort, Vidafel Conversion in other *ejidos*	No major change, yet

Source: Data from the author.

LAND CONVERSION: PRIVATIZATION OF *EJIDO* LAND

This section compares and analyzes land conversion in the three *ejidos* and the reasons for that transformation, which implies land use changes. It is important to note that the three *ejidos* studied in this research were undergoing changes not exclusively owing to the 1992 Reforms, but also owing to the way they developed historically, their location, resources, and the kind of capital investment that influenced the development of their regional economic base. In this sense, Article 27 has formalized the existing informal processes of *ejido* land privatization, but by doing so, has also accelerated privatization in certain *ejidos*.

Of the three *ejidos* studied, only Ixtaltepec fulfilled a traditional rural image. SLRC and La Poza were experiencing a process of accelerating change and did not correspond to that image. On the one hand, SLRC was a rapid-growth *ejido* undergoing urbanization. Transformations in land and labor were taking place, mainly because land was no longer used for agriculture or agriculture-related purposes. In SLRC, land was used both for industry and housing, while in La Poza it was used for tourism, retail, commercial and housing. To a lesser degree, Ixtaltepec was selling land for housing purposes.

The processes of transformation of these *ejidos* were different. In Ixtaltepec, land conversion achieved through privatization started long before the approval of the 1992 *Ejido* Reforms. People migrating to the oil regions or to Mexico City sold their land, rented it, or left it for relatives to land-sit, even when it was not permitted by law. Some other *ejidatarios* died and their descendants who were no longer *ejidatarios,* or did not work the land, sold the land. In any case, people started privatizing their land first by renting it, then subdividing it, and finally by selling it. Land Privatization and land conversion in Ixtaltepec had not been a massive process, rather, it had responded to the in- and out-migration in town. In addition, land conversion had taken place in order to satisfy housing demands from relatives, or people from Oaxaca or Juchitán cities who were buying land in *ejidos* because it was more affordable than in cities.

The principal threat of future privatization in Ixtaltepec came from a proposal by Japanese capital to build windmills in the *ejido*. In order to carry out this project, Japanese investors and the *Procuraduría Agraria,* which is the institution in charge of implementing the Titling and Certification program, were requiring the *ejido* to title their land. Residents of the *ejido* had mentioned that people from Oaxaca City visited the area to measure the *ejido* and to notify them that they would be included in the PROCEDE program. They had not been informed that, in order to start the measuring process, *ejidatarios* needed to approve by consensus their participation in the program.

In comparison to Ixtaltepec, residents in San Luis Río Colorado were familiar with quick changes in land use. Few agricultural lands had been left since several expropriations had already taken place in the *ejido* in order to convert *ejido* land to urban and manufacturing purposes. In fact, the city of San Luis had gradually grown over time to occupy approximately 26 percent of the total *ejido* land. Originally, these expropriations, which had taken place since the 1940s, had the purpose of "regularizing" *ejido* land sold illegally for urban purposes.

San Luis Río Colorado became the first *ejido* to privatize land under the new reforms. Previous to the approval of the 1992 Reforms, international capital and *ejidatarios* were negotiating the construction of an international industrial park on 3,441 hectares of *ejido* land. Since legislation did not allow that move, they were considering expropriation. That is, they were asking the government to expropriate land "in the name of the public utility." The government would compensate *ejidatarios* for their expropriated land. The land would then be sold or given in concession to the investors so that they could build the industrial park. *Ejidatarios* were applying for privatizing that land, even before the approval of the *Ejido* Reforms. As soon as the new reforms were approved, foreign investors and *ejidatarios* saw this policy as an opportunity to undertake privatization. Once the *ejido* privatized its land, it was transferred to an entity representing the - ex-*ejidatarios*, now private proprietors, called COINE.[1] In turn, COINE participated in the Nafta group which was the name of the joint venture between global private capital and ex-*ejidatarios*.

It is evident that in SLRC, the *Ejido* Reforms eased the process of land conversion and subsequent urbanization. Although the Reforms did not originate the process itself, they formalized it. They also opened the flood gates for further land conversion to occur since construction of the industrial park would generate population pressures over housing and services and, consequently, the need for further urbanization. The possibilities of survival of the remaining *ejido* land were scarce.

The transformation of *ejido* land in La Poza has been violently carried out by the State in order for the national and international hotel industry to build tourism resorts.[2] As in SLRC, privatization and conversion of *ejido* land has taken place by means of expropriation, but in La Poza expropriations have been violent. Through expropriations, *ejido* land has been privatized, sold to investors, and converted to non-agricultural uses. Residents of the *ejido* have been relocated in a strip of land behind the original land of the *ejido*. In La Poza, land was expropriated, privatized, and then legally converted to land for tourism development to build the Vidafel project.

The *ejido* land of La Poza has also been privatized and converted through regular sale. The high number of *avecindados* in La Poza show that subdivision and sale of *ejido* land existed previous to the *Ejido* Reforms. *Avecindados* are residents of the *ejido*, who own *ejido* land by purchasing it from the *ejido*. However, by obtaining *ejido* land, they did not obtain *ejido* rights; they were not allowed to vote or to be part of decision-making within the *ejido*. The constant expropriations in the region have made *ejidatarios* think that the rest of their *ejido* land would soon be expropriated and, if they did not secure a title or certificate proving their ownership, the State would again pay minimal prices for their land. In any case, because of their proximity to Acapulco, the *ejido* was quickly becoming part of the city. In this *ejido*, expropriation and *ejido* land privatization fostered urbanization since the relocated *ejidatarios* created new settlements in previously uninhabited areas.

LABOR FORCE TRANSFORMATIONS

The labor force in the *ejidos* is transforming due to migration, waves of investment, and *ejido* land use conversion. These factors have affected the composition of the population in the *ejido* as well as their waged and non-waged occupations. As Article 27 favors land use changes and new waves of investment, it contributes to further labor force transformations, mostly in those *ejidos* that historically have been located closer to cities or those that belong to regions with leading economic activities. The labor transformations observed in the three *ejidos* studied point to gender differences of the labor markets in process of appearing and disappearing. Other elements to be considered in the analysis of the labor markets in the *ejidos* were ethnicity, class, geographic origin, and generational characteristics.

In Ixtaltepec, for example, the male labor force continued to work in agriculture and pottery making. Some of the younger men and women migrated to the oil producing regions, leaving the older population behind, while others migrated to the city of Oaxaca or Mexico City. Women worked in a self-created, informal labor market. They usually traded produce, food, and sometimes pottery as their mothers and grandmothers had done before them. Although they did not work in agriculture as women of older generations had, and they did not own their land, they still had access to resources from the land and were able to trade their products. In order to trade their merchandise they traveled, often for a number of hours or for several days, to nearby towns or throughout the Isthmus region. Their husbands and relatives took care of their children during their absence. Both men and women adjusted their schedules to accomplish their different economic activities.

Men in San Luis Río Colorado owned *ejido* land, but that did not mean that agriculture was their main activity. They had multiple activities; they could work in their *ejidos* or become part of the transnational communities that crossed the border to work in the United States. The service and manufacturing sectors were increasingly expanding in this region due to its proximity to the high-tech region of California, which was relocating and/or opening new high-tech oriented *maquiladoras* on the Mexican side, and especially along the border of Sonora. The women in San Luis Río Colorado who worked in the *maquiladora* industry were mostly young and childless. Older women became housewives and took part in other informal activities. Unlike women in Ixtaltepec, some women in SLRC owned *ejido* land. Many were single mothers and heads of households because they had migrated to this area either by themselves, with their children, or with their husbands who later crossed the border and never returned. Like the male population, some women lived and/or worked on both sides of the border. Some of them held U.S. work permits and a few of them owned property in the United States.

In contrast, the tourism industry in Acapulco was destroying the local labor markets for men and women in coastal *ejidos*. The industry, however, was not replacing the jobs that had been eliminated. Fishermen, farmers, and nursery workers were jobless literally from one day to the next. Tourism also did not provide enough jobs for the women from La Poza.[3] Jobs in tourism were temporary, so, both men and women combined formal and informal activities in order to generate survival incomes.

Women worked in the family-run plant nurseries, watering and weeding plants. They also prepared food and coconut candy to sell. Some of them went to Acapulco to clean houses. A few were employed as maids or cooks for the hotels in the area. When men lost *ejido* land, women were also affected; when they lost their land they were losing not only their economic support, but their housing space, too.

RURAL/URBAN TRANSFORMATIONS IN *EJIDOS*

Over time, the three *ejidos* studied in this paper have undergone different degrees of *ejido* land conversion in which different actors, outside and within the *ejido*, were motivated by various purposes. Privatization and conversion of *ejido* land have restructured land use, *ejido* populations, occupations, labor markets, and the composition of the labor force within the *ejido*. Consequently, they have influenced the way men and women participate in their local economies.

Land conversion has been achieved in several ways—such as illegal sale of *ejido* land for housing purposes; both violent and "peaceful" expropriations of *ejido* land by the State; and, lately, privatization of *ejido* land to organize joint ventures with capital. It is true that appropriation of land for non-agricultural uses was not a process generated only by the 1992 *Ejido* Reforms. Nonetheless, the Reforms accelerated this process by easing up on the legal restrictions to privatization of land and its conversion to other uses. Now that the legal paradigm is in place, *ejido* land privatization and conversion will continue to be encouraged. In spite of the illegal sales of *ejido* land carried out by *ejidatarios*, the structure of the *ejido* land tenure until 1992, which did not allow the alienation of land, functioned as a growth control mechanism that is no longer in place.

Although privatization and land conversion occurred previous to the *Ejido* Reforms, they were mainly accomplished informally between individuals as a way to have access to cheap land for housing. Capital on the other hand gained access to *ejido* land through the expropriation mechanisms implemented by the State whenever land needed to be made available. The current deregulation of land alienation broadens the participation of capital in acquiring *ejido* land and converting it to other uses. In the cases studied in this paper, the oil, the *maquiladora*, and the tourism industries have historically accumulated land. Under the protection of the *Ejido* Reform, owners of capital are able to accumulate further *ejido* land and create land monopolies. This is already the case in the *maquiladora* and tourism industries that have become border *maquiladora latifundios* or coastal tourism *latifundios*.

Land privatization has restructured land uses in *ejidos* and has given a different character to the regions to which they belong. From the three *ejidos* studied here, Ixtaltepec is the one that has undergone the least transformation in land use. Still, the in- and out-migration generated by the impact of the oil industry in the region has provoked subdivisions and conversion of agricultural land for housing purposes. If the wind mills that Japanese capital wants to build are approved, further land conversion is probable. The earlier land conversions, however, did not radically alter the local trade economy since men continued working in agriculture and women, in spite of not owning the land, had continual access to produce and products through their relatives and partners who did own land.

This was not the case in San Luis Río Colorado where changes in land use affected the composition of the population. That is, the *maquiladora* industry attracted workers from other states who demanded land for housing and services. The *maquiladora* industry itself converted land from agricultural into manufacturing uses. Men, and especially women, who were once devoted to agriculture or were of rural origin, gradually became *maquiladora* workers. The women's labor market experienced a generational change because the *maquiladora* hires mostly young women. While older women used to work in agriculture or agriculture related activities, younger women are now working for a wage in *maquiladoras*. The international industrial park built in the *ejido* converted more *ejido* land to manufacturing uses and reinforced the region as *maquiladora* territory. The privatized *ejido* land consisted of parcels that were not used for other economic activities; however, its conversion increased pressures on *ejido* land as the international industrial park contemplated the construction of a new customs office. Additional land for infrastructure, housing, and services will certainly be needed.

In comparison to Ixtaltepec, more men and women in SLRC are owners of land in nearby *ejidos*. These populations participated in the organization of *ejidos* and obtained the land in spite of not having worked directly in agriculture. Currently, some of them rent out their parcels, a few hire laborers to work their land, and others keep the land as a potential resource. Since SLRC is quickly urbanizing, they will have the possibility to sell it for urban or other non-agricultural purposes.

Land conversion in La Poza, Guerrero, has had an impact on men's and women's occupations by generation. For years, they worked in agriculture and coconut production and later plant nurseries, which are family-run businesses. In a way, plant nursing is a survival strategy, since periodic expropriations of land have decimated the size of parcels and made them too small for agriculture or coconut production. Where land was expropriated, nurseries were often destroyed and the affected populations no longer work in this sector.

Urbanization in the city of Acapulco and tourist developments in nearby *ejidos* have prompted migration into La Poza, and *ejidatarios* have subdivided and sold their land. Land conversion and urbanization have changed the composition of the population as in SLRC. The difference is that in La Poza, *ejido* land is converted for tourism purposes, while in SLRC the land is for manufacturing purposes.

Although the plants grown in nurseries are destined for the tourist industry in the region, the *ejido* land for those nurseries is decreasing. As a rule, tourist resorts do not provide jobs for the *ejido* population. However, when exceptions are made, the resorts hire only young educated men and women, which is a small percentage of the *ejido* population. In addition, jobs in this industry are mostly temporary.

In La Poza both non-paid and wage labor coexist, the first in the form of family labor for the nurseries, the second as temporary jobs for the tourism industry as well as other paid activities in Acapulco City. Some of the population that own land in La Poza are not *ejidatarios*, but *avecindados*, which means that they migrated to the *ejido* and bought land for housing purposes. Like *avecindados* in SLRC, *avecindados* in La Poza do not work the land. Some *avecindados* have built their houses on *ejido* land because it was cheaper than in the city, and a few of them own the land as a real estate asset; actually, this latter group is really not *avecindados*, as they do not reside within the *ejido*.

In addition to the Vidafel resort that displaced *ejidatarios* in La Poza, the tourism industry is planning new expropriations.[4] In the near future, residents of La Poza may be displaced and relocated to areas not suitable for nursery production, and consequently their occupations may change again.

ASSESSING ARTICLE 27 IN THE THREE *EJIDOS*

Mexico, a country comprised of more than 25 states, has a myriad of regional and cultural differences. Examples of these profound differences are the Northern and Southern regions that economically have developed unevenly and, culturally, have defined different Indigenous and non-Indigenous populations. To describe it in terms of the development jargon that racialized regions and populations, the North traditionally has been more "developed" and culturally mixed, and the South more "underdeveloped" with more Indigenous groups. Within this framework, the approval and application of the 1992 Article 27 in different regions resulted in greater unevenness as global capital chose to locate in certain regions, further exacerbating their economic differences.

Critics of the 1992 Article 27 claimed that the *Ejido* Reform was another privatization policy that supported the foundations of the neoliberal model in Mexico and the globalizing economy. They argued that *latifundios*, in other words land monopolies, would form as impoverished *ejido* owners sold their land and migrated out. As part of this cycle, they emphasized that rural migration to the United States would increase as peasants were alienated from their lands. On the other hand, supporters of the *Ejido* Reform claimed that the new legislation would bring justice and economic equity to rural Mexico. Both critics and supporters of Article 27 focused on the impacts of this policy on agricultural and rural areas, but failed to acknowledge that not all *ejidos* are rural and that their local economies are not based exclusively on agriculture.

The debate over the impacts of Article 27 regularly omitted the urban, regional, and global implications of the policy. For some, privatization of *ejido* land was the main concern, but privatization was already taking place in rural and semi-rural Mexico. *Ejidatarios* were selling their lands or part of them for different reasons. Some of them subdivided it and sold it to relatives for housing purposes. Others sold it completely out of necessity and left the countryside. In other cases, the State expropriated *ejido* land. In any case, land privatization is not new. The new element is the speed with which this privatization is taking place and the celerity with which land is being converted to non-agricultural uses. In this sense, it is worth repeating that *ejidos* were a growth control mechanism that is no longer in place.

The haste in applying Article 27 and the promptitude of investors in securing *ejido* land did not translate into a more efficient way of informing the communities affected. During my field research, I learned that most of the *ejidatarios* I interviewed did not know what the new land legislation was proposing nor did they understand the implications of such a policy. According to PROCEDE personnel, in SLRC, the process used to privatize land for the international industrial park did not follow the process established by law. In Ixtaltepec, precisely when I was leaving the *ejido*, the announcement was made of the privatization and remodeling of the Transisthmian railroad, which runs from Salina Cruz, Oaxaca to Coatzacoalcos, Veracruz. This railroad, along with a superhighway and two energy-generating plants will attempt to "make the Isthmus of Tehuantepec compete with the Panama Canal."[5] In order to pursue those projects, *ejido* and communal lands in the region would require privatization and conversion. In fact, one of those energy-producing plants may be located in Ixtaltepec, which may prompt future land conversion. However, in Ixtaltepec, the plans to privatize land for those energy-producing plants were unknown by most *ejidatarios*. Finally, in La Poza, the right of *ejidatarios* to land was overlooked when their lands were violently expropriated for tourist purposes. Information was not provided about the new legislation and, in some cases, coercion was used to obligate *ejidatarios* to participate in the PROCEDE.

With the exception of Ixtaltepec, the *ejidos* of SLRC and La Poza are very different from the traditional *ejidos* with a rural character or with an agriculture-oriented economy. These *ejidos* still undertake some agricultural activities and have preserved some rural spaces, but they have been deeply altered by national and global capital investment and urbanization and no longer belong to the traditional typology of rural *ejidos*. These are *ejidos* in transition towards a semi-urban or urban image, mainly because of their proximity to urban areas and their receipt of national and transnational capital.

All three *ejidos* have experienced diverse privatization and land use conversion processes over time. These processes have affected their traditional economic activities. In the case of Ixtaltepec, Oaxaca, privatization and land use conversion were still controlled by the *ejido* authorities; thus, no major impact was felt on the labor force since they continued their traditional occupations. In SLRC and La Poza, however, *ejido* land privatization and conversion carried out by the cooperation of the Mexican State and national and transnational capitals contributed to diminishing agricultural activities. In these two latter *ejidos,* the presence of the *maquiladora* and the tourism industry, respectively, was reinforced. Some residents of the *ejido* were employed in those industries, but jobs created by those industries were mostly taken by immigrants from other areas.

Participation of the female labor force in Mexican *ejidos* has always been present. The difference today resides in the kind of work women do, compared to what they used to do, and the pay they receive. Most of the women interviewed for this study had an agricultural or agriculture-related activity, as in La Poza or in SLRC. However, women did not receive wages while working in those sectors. As regions changed because of the investment received and by in-migration generated by the establishment of new industries, some of the women's activities changed and they began obtaining salaried jobs, as was the case in SLRC.

Along with the establishment of the *maquiladora*, tourist, and oil industries have transformed land, labor, and local economies, and have reshaped the character of the regions to which those former *ejidos* belonged. Often, the relocating industries did not hire residents of the locality as was the case in the Vidafel project in La Poza. For this project, Vidafel hired mostly people from Acapulco City and other tourist cities, but not from the *ejido* or rural areas.

Ejido land conversion into urban, manufacturing, or tourist land indicates a transition from a rural into an urban environment. This new setting also implies a transformation in the gender relations within the family. In Ixtaltepec and La Poza, household arrangements took various forms, while in SLRC the disintegration of families and the increase in the number of single mothers displayed a disruption in traditional family arrangements. This is a topic for future research, as the consequences of *ejido* land privatization regularly show the value of the economic benefit for capitalists, but not the economic and social disruption of families and communities.

Land tenure in *ejidos* is also changing. Land concentration is under way. Eighty four years after the Agrarian Reform attempted to break up *latifundios*, new *latifundios* are reappearing, not as haciendas, but in the form of tourist resorts and *maquiladora* factories or industrial parks. Contrary to their promises for more jobs, these developments are disrupting the existing local economies and are not providing enough jobs for the dislocated labor force, both male and female. Nor are they providing housing.

The Reforms to Article 27 are fostering processes of urbanization in some *ejidos*. In the three *ejidos* studied here, urbanization takes place unevenly. In Ixtaltepec this process is at a very incipient stage, while La Poza and San Luis Río Colorado are rapidly converting *ejido* land into manufacturing, tourism, and housing uses. These latter two *ejidos* are no longer rural areas. San Luis Río Colorado is a city and La Poza is merging into Acapulco City.

These transformations in *ejidos* might have created opportunities for a few, but have cut off opportunities for many, especially for those originally residing in the local *ejido*. These transformations have destroyed local economies, as was the case in SLRC and La Poza, and have given birth to temporary labor markets requiring a flexible labor force to operate *maquiladoras* and work for the tourist industry. The labor force preferred by these industries consists of young and childless women. The requirements of cheap labor and cheap land by capital have been met successfully. The most affected populations in this process are those men and women whose jobs for supporting a family are being systematically destroyed or whose possibilities to integrate in the newly created labor markets are meager.

CONCLUSIONS

In 1998, the report of the *Procuraduría Agraria*—the institution in charge of carrying out the titling and certification process—declared that by that year 21,756 out of the 27,144 *ejidos* existing in Mexico had decided, in their internal assemblies, to participate in PROCEDE, which accounted for 80 percent of the total *ejidos*. From these, about 60 percent, or 16,175 *ejidos*, had concluded the final phase of certification, which, according to the report, implied the "regularization" (as if they had been irregular before) of 36.6 million hectares that included 1.9 million *ejidatarios*.[6]

The report indicated the regional differentiation in applying PROCEDE, pointing out the *ejidos* in Central and Southern Mexico as those lagging in completing the certification and titling program. Map 8.1 clearly displays this regional differentiation. About 75 percent of *ejidos* in small states such as Tlaxcala and Morelos in Central Mexico, along with Colima on the Pacific coast, and Aguascalientes in the Central North have been certified by PROCEDE.[7] In comparison to the Northern region, where 56 percent to 74 percent of *ejidos* have been certified, just 33 percent to 54 percent of *ejidos* in the South have participated in the program.

In addition, in its section titled *Programa de Incorporación de Suelo Social* (PISO), or Program of Incorporation of Social Land, the report mentions the conversion of "social land," that is *ejidos*, into private property in order to "satisfy the demands of housing, urban infrastructure, and regional development." In this document, the "modernization of the Mexican agriculture" was omitted; instead this new PISO program was launched specifying that "the National Program of Urban Development proposes to

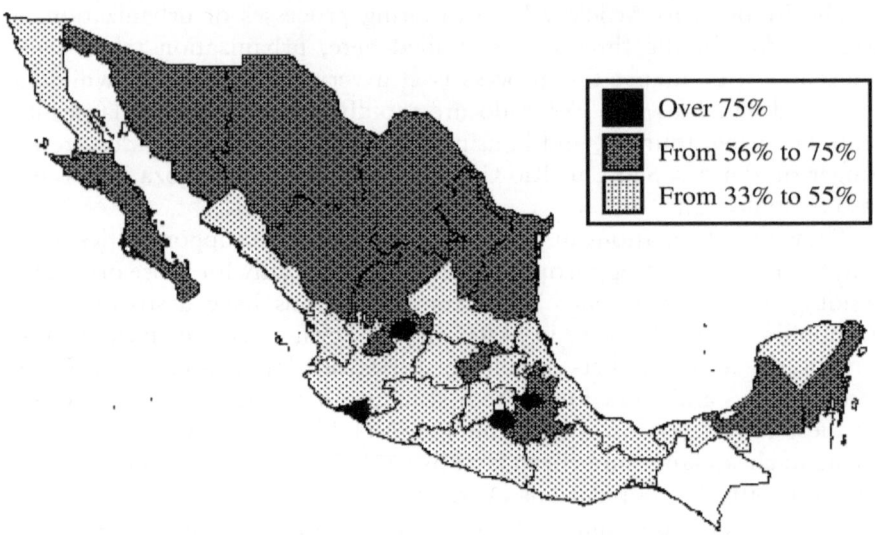

Over 75%

From 56% to 75%

From 33% to 55%

Map 8.1 Application of PROCEDE by State and Region, 1997. *Source:* Guillermo R. Zepeda Lecuona, "Cuatro Años de PROCEDE: Avances y Desafíos en la Definición de Derechos Agrarios en México," *Estadios Agrarios Revista de la Procuraduría Agraria No. 9* (np, October 1997–April 1998), 6.

incorporate by year 2000 about 150,000 hectares of land,"[8] of which 100,000 hectares were *ejido* lands. The methods listed to "incorporate" this land were: **privatization through expropriation, privatization to create mercantile societies,** and regular **privatization.**[9] Since the cases of the *ejidos* La Poza in Guerrero, San Luis Río Colorado in Sonora, and Ixtaltepec in Oaxaca demonstrate, respectively, those methods of privatization of *ejido* land, I can state that the cases presented in Chapters Five through Seven of this research illuminate the beginnings of a new stage of land privatization in Mexico, the privatization of *ejidos* in the age of NAFTA.

In the same vein, the report states: "Based on this [program] and on the 100 Cities Program,[10] it has been proposed the incorporation of 819 agrarian nuclei, of which 411, slightly more than 50 percent, have culminated their certification process; therefore, they are qualified to participate in urban development projects."[11] The urban aspect of the 1992 Article 27 Reforms is openly acknowledged again when the same report specifies: "The Procuraduría Agraria intends that in 1998, through PROCEDE, *ejidos* design an alternative of regularization, according to the urban development plan of the municipality in which they are located."[12] Supposedly, this municipal plan includes *ejidatarios* in the planning process of the urbanization of *ejidos*. The proposal sounds fair, but ignores the internal and external dynamics of *ejidos* and the different actors that participate in their use conversion. It ignores the authoritarian planning through which the *Ejido* Reforms came into place and which would prevent the real participation of *ejidatarios* and *ejido* populations in planning the urban future of *ejidos*. It ignores the fact that this policy was also pushed by supranational institutions that now are backing off from privatization policies of communal lands.

In the July 1998 World Bank's Land Policy Report, entitled *The Evolution of the World Bank's Land Policy*, this supranational institution qualifies the titling programs that were once promoted as "narrow interventions" and acknowledges "significant changes in the World Bank's policy recommendations."[13]

Although the different uprisings and protests organized by Indigenous and other rural communities are not mentioned as causes of this change in policy recommendations, it is evident that, at least in the case of Mexico, the Zapatista uprising in the Southern state of Chiapas brought attention to the fact that land policies designed top down were just not good for the Indigenous communities living in Mexico. The Zapatista movement, that many called "regional and local," was a response that many of the Indigenous communities around the world strongly supported. Could that be a factor in the change of policy of the World Bank? Was the increasing impoverishment of the countryside in those countries that had applied structural adjustment and land privatization programs also a factor? In any case, the changes included in this report specify:

> Instead of recommending the abandonment of communal tenure systems in favor of freehold title and subdivision of the commons (as in the LRPP),[14] the Bank's now recognizes that communal tenure systems are often a more cost-effective solution than formal title and that, in situations where this is the case, efforts to reduce the cost of operation, improve accountability, and facilitate evolution of communal systems in response to local needs may be needed.
>
> In addition to clarifying the preconditions for titling to be economically viable, research has shown that in situations where credit markets are imperfect, the benefits from titling programs can easily favor the better off, implying a need for greater attention to the functioning of other factor markets and the equity impact of titling programs.
>
> Instead of expecting liberalization of land sales markets to lead to instantaneous transfers of land to more efficient producers, it is now understood that under (credit) market imperfections common in rural areas or in the presence of policy distortions, lands sales markets may increase neither efficiency nor equity and a more integrated approach to the development of rural factor markets is needed.[15]

Similarly, a year later, in 1999, a paper prepared for the International Monetary Fund's Conference on Second Generation Reforms[16] stated that "land titling has not led to significant investments in agriculture" or in productivity.[17] Rather, in several instances, it has created "uncertainty and conflict over land rights,"[18] gender inequality, and insecurity and "manipulation and abuse by elites."[19] One would think that borrowing countries could legally take action against the policy "mistakes" that supranational financial institutions have inflicted on them; however, the lending agreements that those countries have signed with the institutions highlight that no legal action can be taken against those institutions. Immunity for these institutions is part of the deal.[20] Finally, the paper acknowledges "the flexibility of indigenous tenure systems" to meet "the needs of the landless and commercial farmers."[21] In spite of the acknowledgment that both the World Bank and the International Monetary Fund have publicly made, PROCEDE continues in Mexico.

Twelve years after Article 27 was reformed and eleven years after PRO-CEDE was initiated, land use changes have continued to occur in the Mexican landscape. SLRC has not been the only *ejido* that privatized its lands; since then 12 other *ejidos* followed such a privatization path, and another 22 were in the process of constituting real estate mercantile societies like that of SLRC.[22] Other mega-projects have been proposed and initiated in *ejido* lands. In spite of the recognition on the part of the supranational institutions that "privatization of property rights is clearly not a magic solution,"[23] and that "[p]roperty rights are a common good,"[24] the effect of the Reforms is now felt in Mexico as urbanization has increased, regional differences have exacerbated, and global capital continues to find *ejido* space in which to relocate. It is obvious that when structural adjustment policies claimed that "communal tenure arrangements have often been considered an economically 'inferior' arrangement,"[25] they did not make clear that what they meant by "inferior" represented an obstacle for the relocation of global capital. That "obstacle" has been removed.

Along with that "obstacle" historical rights and the memory of Indigenous struggles over their territories were removed. If law can be changed at the will of the national and transnational elites, this means that the notion of rights has become "flexible." Almost overnight, the rights of landless and Indigenous communities to their lands have given way to the right to "security" for the investor.

By January 2001, the Fox administration admitted that 48 percent of *ejido* land remained to be certificated.[26] Thus, the then Secretary of the Agrarian Reform, Teresa Herra Tello, indicated that PROCEDE would continue because "the *ejido* and the community must adapt their functioning to the new demands of efficiency, competitiveness, and innovation that the global world imposes."[27] She indicated that her agency would promote the creation of new mercantile societies to carry out "ambitious projects" in agriculture, real estate, and tourism.[28]

These "ambitious projects" harmonize with a recent proposal formulated by the Mexico City Association of Architects and the Institute of Economic Research of the National Autonomous University of Mexico (UNAM). This proposal was requested by the *Secretaría de Desarrollo Social* (SEDESOL), or Secretary of Social Development, and its objective was to "visualize" the urban and regional future of Mexico.[29] The document visualizes nothing new. It proposes more economic support to the North and not the South, except for tourist and oil-producing regions.[30] San Luis Río Colorado is among the Northern cities classified as "with high capability to support the national development,"[31] while "[i]n the South and Southeast regions, no city of prime level is located."[32] However, Acapulco is classified as a city with "lower-middle capacity." According to the report, these cities are located in the "Central regions of the country."[33] Finally, the report states that the "cities with low capability" are "notoriously located in the Center and South of the Pacific Ocean." According to this visualizing classification, Juchitán would be one of these low-capability cities. Since Ixtaltepec is located in the municipality of Juchitán, the same description would apply to the *ejido*.[34] But these classifications imply national policies and economic support which indicates once again that urban and regional policies will continue to increase uneven development in Mexico and aggravate the conditions of social inequity and injustice in the South.

Since the *Ejido* Reforms are still under way, future research needs to focus on the consequences of further privatization of *ejido* lands, and on the increase in uneven development as well as on divergent class and ethnic interests both within and outside the *ejido*. During the years that PROCEDE has been applied, it generated a wealth of data that before 1992 were not available. These data need to be analyzed and exposed in order to shed further light on the internal organization of *ejidos*; the way titling and certification was accomplished; the cases that the Agrarian Tribunals had to resolve; and the trends that *ejidos* are following in terms of land privatization, further urbanization, regional development, and the relocation of new economic activities. Most significantly, these data should be used to inform land policies and planning policy.

The current discussion on the Indigenous Law in Mexico has to necessarily be part of this future research as well as of the current design of land policy and planning. This inclusion and recognition of Indigenous communities is urgent, as land, land use, and urban and regional planning in Mexico has continuously excluded—and continues to exclude and dispossess—Indigenous peoples, both in *ejidos* and agrarian communities, as well as in the urban settings they later occupy. The historical exclusion and dispossession of Indigenous peoples are well described by Commander Javier from the Zapatista Army who said, "to be Indigenous means that I am not recognized by the [Mexican] federal and state laws,"[35] or in the words of Comandante Maxo, "We feel sad because the government is taking away all our patrimony."[36]

In the discussions of land policy and planning, the narrow definition of Indigenous that the Mexican government has coined should change. Indigenous peoples are now in cities; even when they do not speak Indigenous languages, they still claim their heritage and are proud of their background. That is why, during the demonstrations in downtown Mexico City in support of the Zapatistas in Chiapas, people shouted the slogan of "We all are Indians," or as subcommander Marcos put it: "To be Indigenous in today's Mexico, means to fight for respect and dignity for all of those who are excluded and despised. It means to fight for the Indigenous people, but also for the women, for the youth, for the children, for the homosexuals and lesbians, for the disabled, for the elderly, in short, for all the different ones."[37] Many of those different ones in today's Mexico are urban Indigenous people, or their descendent, who have been historically displaced from their lands, from their spaces.

Thus, in spite of land privatization, urbanization, uneven regional development, and globalization processes, the identity and the history of many Indigenous communities—be they rural or urban, be they close or far away—are still attached to land and its culture. History, memory, rights, and identity must inform planning for social justice; otherwise, authoritarian planning will continue to exclude the majority of the population of Mexico, many of whom have recognized that "all of us are Indians."

Appendix

1. Name	2. Age
3. Place of origin	4. Marital Status
5. Scholarship	6. Number of Children

7. Age, sex, scholarship, occupation, and place of residence of sons, daughters, and partner.

Age	Sex	Scholarship	Occupation	Place of residence

8. How many people live in your house?

9. Age, sex, family relationship and occupation of all the people who live in your house.

Age	Sex	Family Relationship	Occupation

10. Place of origin and current occupation of parents and partner.

	Place of Origin	Occupation
Mother		
Father		
Partner		

11. What were the income generating activities that your parents and partner used to do and they do not do anymore?

	Activities
Mother	
Father	
Partner	

12. Place of origin and occupation of maternal and paternal grandparents.

	Place of Origin	Occupation
Maternal grandmother		
Maternal grandfather		
Paternal grandmother		
Paternal grandfather		

13. What income generating activities did your maternal and paternal grandparents used to do that they do not do anymore?

	Activities
Maternal grandmother	
Maternal grandfather	
Paternal grandmother	
Paternal grandfather	

14. What is your occupation?

15. What time do you get up and what time do you go to bed?

16. What are your activities during the day and how many hours do you spend in each activity?

Activity	Number of Hours

17. Who participates in the house chores?

18. What chores do your partner and the rest of your household do at home?

Person	Activity	Number of Hours

19. What age did you get married?

20. What kind of income-generating activities did you do before getting married?

Activities	

21. Do you use wood to cook? Yes___ No___
If you do, who is in charge of bringing wood to home?

22. Do you have drinking water at home? Yes___ No___
If not, who brings water to home?

23. What activities did your partner do before getting married?

Activities	

24. How many brothers and sisters do you have?
    Mention their age, sex, scholarship, occupation and place of residence.

Age	Sex	Scholarship	Occupation	Place of residence

25. Who contributes to the household income?

26. Who is in charge of administering the household income?

27. Do you know how much your partner makes? Yes__ No__

28. Does your partner know how much you make? Yes__ No__

29. Who decides about money when there are big expenses to make?____

30. Do you own your *ejido* land? Yes__ No__

31. How did you obtain your *ejido* parcel?
    Sale_____ Inherited_____ Awarded_____

32. In case you bought *ejido* land, who did you buy it from?____

33. Have any women in your family inherited *ejido* land? Yes__ No__

34. Do you work in agriculture? Yes__ No__ Sometimes_____

35. Do you participate in *ejido* meetings? Yes__ No__ Sometimes_____

36. Is it accepted in the *ejido* that women work outside home? Yes__ No__

37. Do you know anything about the modifications to Article 27 of the
    Constitution? Yes__ No__

38. Has PROCEDE been implemented in this *ejido*?
    Yes__ No__ When?____Why not?____

39. Have you received funds from PROCAMPO?
    Yes__ No__ Why not?____

40. How did you learn about PROCAMPO? Through Friends____
    Through the Procuraduría Agraria____ Other____

41. Was it easy to obtain that funding?
    Yes__ No__ Why?____

Notes

NOTES FOR CHAPTER ONE

1. Charles C. Cumberland, *The Meaning of the Mexican Revolution* (Boston: D.C. Heath and Company, 1967), vii.
2. *Latifundios* are large land holdings.
3. The populations that work and own an *ejido.*
4. Procuraduría Agraria, *¿Qué es el PROCEDE?* (México, D.F.: Procuraduría Agraria, 1993).
5. For further reference, see the *Transformation of Rural Mexico Series* that published the results of the *Ejido Reform Research Project* sponsored by the University of California, San Diego during the period 1992–1997.
6. In comparison to the San Diego school, the outreach of the Austin school was limited and no major publications resulted from their meetings. For further reference, see the papers resulting from the conference, *The Urbanization of the Ejido: The Impact of the Reform to Article 27 Upon Real Estate Development and Land Regulation Policies,* that took place February 4–5, 1994 at the University of Texas at Austin.
7. In 1998 the book edited by Wayne A. Cornelius and David Myhre, *The Transformation of Rural Mexico: Reforming the Ejido Sector,* included a chapter by Gareth A. Jones and Peter M. Ward entitled "Deregulating the Ejido: The Impact on Urban Development in Mexico" (San Diego: Center for U.S.-Mexican Studies, University of California, 1998), 247–275.
8. This debate will not be analyzed in this research, for your reference, see Cumberland, *The Meaning of the Mexican Revolution,* 1967.
9. North America Free Trade Agreement.
10. I was a grantee of the UC-San Diego *Ejido* Project and kept updated about their different publications. I also attended the UT Austin conference and collected the papers presented in that conference.

11. Bryan Roberts, "The Place of Regions in Mexico" in Eric Van Young, ed., *Mexico's Regions: Comparative History and Development* (San Diego: Center for U.S.-Mexican Studies, University of California, 1992).

12. Edward W. Soja, *Thirdspace: Journey to Los Angeles and Other Real-And-Imagined Places* (Cambridge, MA and Oxford, UK: Blackwell Publishers, 1996).

13. Roberts, "The Place of Regions in Mexico," 230.

14. Henri Lefevre, *The Production of Space,* translated by Donald Nicholson-Smith (Oxford, UK and Cambridge, MA: Basil Blackwell, 1991).

15. Soja, *Thirdspace,* 2.

16. Ibid., 3.

17. Ibid., 17.

18. Billie R. DeWalt and Martha W. Rees, *The End of Agrarian Reform in Mexico: Past Lessons, Future Prospects*, (San Diego: Center for U.S.-Mexican Studies, University of California, 1994), 4.

19. Robert Yin, *Case Study Research: Design and Methods* Second edition (Thousand Oaks, London and New Delhi: Sage Publications, 1994), 3.

20. R.E. Stake, The Art of Case Study (Thousands Oaks, CA: Sage, 1995).

21. The questionnaire used to interview families in *ejidos* is listed in Appendix A.

22. Norman Denzin and Yvonna Lincoln, *Handbook of Qualitative Research*, (Thousand Oaks, London and New Delhi: Sage, 1994), 2.

23. See Chapter Three to review the functions of these administrative entities.

24. "This technique requires that an initial set of appropriate informants be located through some reasonable means and surveyed. Then they are asked to identify other informants whom they believe are knowledgeable about the matter at issue. These informants are then contacted and asked, in turn, to identify other." Peter Rossi, Howard Freeman and Mark Lipsey, Evaluation, A Systemic Approach (Newbury Park, CA: Sage Publications), 149.

25. I thank Professor Betty Deakin for pointing out the importance of regional differentiation in the incipient elaboration of my research idea while I was still a Master student at UC-Berkeley.

26. Margarita Ordaz Mejía from the Universidad Americana de Acapulco and Alfonso Guzmán Andrade, planner from the City of Acapulco, directed and introduced me to key people in the city. Their support eased my fieldwork in the Mexican state considered the poorest and most violent.

27. Stephen Joseph, planner of the state of Sonora, provided insightful information about *ejidos* in transformation, while Chema Gamboa from the PROCEDE and the economist Daniel Coss Rangel arranged several visits to different *ejidos* and meetings with planners, activists and researchers.

28. Daniel Coss Rangel highlighted the importance of doing field research in border *ejidos* over those surrounding Hermosillo City.

29. For an explanation of communal lands, see Chapter Two.

30. Nemesio Rodriguez from PROCEDE and Rosaria Pisa provided key information about *ejidos* near Juchitán City.

31. Radio and TV ads promoted PROCEDE emphasizing its "voluntary basis". To be a valid process, it required 75 percent attendance of *ejidatarios* to three

required meetings. For more on PROCEDE, see Chapters Two and Three of this research.

32. Roberts, "The Place of Regions in Mexico," 230.
33. See Lynn Stephen, *Viva Zapata!: Generation, Gender, and Historical Consciousness in the Reception of Ejido Reform in Oaxaca*, The Transformation of Rural Mexico No. 6, *Ejido* Research Project (San Diego: Center for U.S.-Mexican Studies, University of California, 1994); Monique Nuijten, *In the Name of the Land: Organization, Transnationalism, and the Culture of the State in a Mexican Ejido* (The Hague, Netherlands: Cip-Data Koninklijke Bibliotheek, 1998); Verónica Vázquez García, ed., *Género, Sustentabilidad y Cambio Social en el México Rural* (Estado de México, México: Colegio de Postgraduados, 1999); Josefina Aranda, Carlota Botey and Rosario Robles, *Tiempo de Crisis, Tiempo de Mujeres* (Oaxaca, México: Universidad Autónoma Benito Juárez de Oaxaca, Centro de Estudios de la Cuestión Agraria Mexicana, 2000).
34. *Mestizos* are mixed populations of Spanish and Mexican ethnicity and culture.

NOTES FOR CHAPTER TWO

1. Charles C. Cumberland, *The Meaning of the Mexican Revolution*, vii.
2. As mentioned in Chapter One, *Mestizos* are mixed populations of Spanish and Mexican ethnicity and culture
3. Jorge Luis Ibarra Mendívil, *Propiedad Agraria y Sistema Político en México* (México: El Colegio de Sonora and Miguel Ángel Porrúa, 1989), 82–83.
4. Ibid., 84–85.
5. Ibid., 86.
6. *Haciendas* were institutionalized estates that controlled both land and labor.
7. Othón de Mendizábal, "Origen de Nuestras Clases Medias" in *Ensayos Sobre las Clases Sociales en México* (México: Nuestro Tiempo, 1970), 14, quoted by Roger Bartra, *Estructura Agraria y Clases Sociales en México* (México, D.F.: Ediciones ERA and Instituto de Investigaciones Sociales de la Universidad Nacional Autónoma de México, 1974; repr. 1991), 119.
8. Bartra, *Estructura Agraria*, 111–112.
9. Ibarra Mendívil, *Propiedad Agraria*, 90.
10. Jesús Silva Herzog, *El Agrarismo Mexicano y la Reforma Agraria* (México: Fondo de Cultura Económica, 1980), 70–73 and 78–82, quoted by Ibarra Mendívil, *Propiedad Agraria*, 91, footnote 25.
11. Frank Tannenbaum, *Mexico: The Struggle for Peace and Bread* (New York: Borzoi Books published by Alfred A. Knopf, 1951), 139.
12. The Porfirio Díaz dictatorship prolonged from 1876 to 1911.
13. Tannenbaum. *Mexico: The Struggle for Peace and Bread*, 139.
14. Friedrich Katz, "The Agrarian Policies and Ideas of the Revolutionary Mexican Factions Led by Emiliano Zapata, Pancho Villa, and Venustiano Carranza" in Laura Randall, ed., *Reforming Mexico's Agrarian Reform* (Armonk, NY: M. E. Sharpe, 1996), 26.

15. Ibarra Mendívil, *Propiedad Agraria*, 95.

16. Ibid., 93.

17. Tannenbaum, *Mexico: The Struggle for Peace and Bread*, 139–141.

18. Salomón Eckstein, *El Ejido Colectivo en México* (México: Fondo de Cultura Económica, 1984), 24–25.

19. Tannenbaum. *Mexico: The Struggle for Peace and Bread*, 140.

20. Ibid., 141.

21. William C. Thiesenhusen, "Mexican Land Reform, 1934–91: Success or Failure?" in Laura Randall, *Reforming Mexico's Agrarian Reform*, 36.

22. Tannenbaum. *Mexico: The Struggle for Peace and Bread*, 144.

23. Ibid., 147.

24. Ibid., 146.

25. Francisco I. Madero occupied the presidency from 1911 to 1913.

26. Zapata (1879–1919).

27. Adolfo Gilly, *Interpretaciones de la Revolución Mexicana*, first repr. (México: Universidad Nacional Autónoma de México [UNAM] and Nueva Imagen, 1985), 22.

28. Victoriano Huerta was in power from 1913 to 1914.

29. Venustiano Carranza was president from 1917 to 1920.

30. Francisco Villa (1878–1923).

31. Katz, "The Agrarian Policies," 22.

32. Ibid., 24.

33. Populations of Spanish–descend born in Mexico.

34. Katz, "The Agrarian Policies," 25.

35. Ibid., 26.

36. Ibid., 26.

37. Ibid., 27–29.

38. Ibid., 31.

39. James Rudolph, *Mexico: A Country Study,* Foreign Area Studies (Washington, D.C.: American University, 1985), 211.

40. Ibid., 212.

41. Ibid., 210.

42. Miguel Acosta Romero and Genaro David Góngora Pimentel, *Constitución Política de los Estados Unidos Mexicanos* (México: Editorial Porrúa, 1984), 288.

43. Rudolph. *Mexico: A Country Study*, 211.

44. Harry Wright, *Foreign Enterprise in Mexico: Laws and Policies* (Chapel Hill: The University of North Carolina Press, 1971), 113.

45. Ibid., 57–58.

46. This regulation included the Church as an entity, but not the individuals working in the Church system.

47. Wright, *Foreign Enterprise*, 117.

48. Ibid., 117.

49. Acosta Romero and Góngora Pimentel, *Constitución Política*, 288.

50. *Caciques* are political bosses.

51. Tom Barry, ed., *Mexico: A Country Guide* (Albuquerque, N.Mex.: The Inter-Hemispheric Education Resource Center, 1992), 154–159.

52. Ibarra Mendívil, *Propiedad Agraria*, 199.

53. Rosario Robles and Julio Moguel, "Agricultura y Proyecto Neoliberal," *El Cotidiano* (México), April 1990, 3, quoted by Barry, *Mexico: A Country Guide*, 361, footnote 80.

54. Carlos Salinas de Gortari was president from 1989 to 1994.

55. Ernesto Zedillo was president from the period 1994–2000.

56. PROCEDE is the Certification and Titling Program for the *Ejido* Lands, further explanation can be found in the section "The 1992 Reforms to the *Ejido* and to Article 27" in this chapter.

57. Later that year, the figure was corrected to 29,474 total *ejidos*. See Miguel Ángel Vidaurri, "Principales Resultados del Censo de Órganos de Representación de Ejidos y Comunidades," *Revista de la Procuraduría Agraria* 10, (May-December 1998) at htpp://200.38.178.160/publica/pa071007.htm.

58. Procuraduría Agraria, *Informe Anual de Actividades*, (México: Procuraduría Agraria, 1998) in web page of the Procuraduría Agraria at http://www.pa.gob.mx/publica/pa07c001.htm#F01.

59. Héctor Manuel Robles Berlanga, "Tipología de los Sujetos Agrarios Procede," *Revista de la Procuraduría Agraria* 4, (July-September 1996) at htpp://200.38/178/160/publica/pa/pa070403.htm, 5–7.

60. PROCAMPO stands for *Programas de Apoyo Directo al Campo* or Programs of Direct Support for the Countryside.

61. "PROCAMPO," *Programas de Apoyo* in the web page of the Secretaría de Agricultura, Ganadería, Desarrollo Rural, Pesca y Alimentación (Secretary of Agriculture, Cattle Raising, Rural Development, Fishing and Food) at http://www.sagarpa.gob.mx/sagar2.htm.

62. Gil Olmos, "Abstencionismo, Seguro Vencedor en Comicios," *La Jornada* (México), 28 September 1996.

63. Herminio Blanco Mendoza, "Investment Implications: NAFTA," *Investing in Mexico*. Report No. 999 (New York: The Conference Board, 1992).

64. Procuraduría Agraria, *Nueva Legislación Agraria*, 2d ed. (México, D.F.: Unidad de Comunicación Social de la Procuraduría Agraria, 1993), 9.

65. Wesley R. Smith, "Salinas Prepares Mexican Agriculture for Free Trade," *Backgrounder* #914 (Washington, D.C.,: The Heritage Foundation), October 1992.

66. Embassy of Mexico in Washington, "Background Information: The Legal Proposal for Mexico's Agricultural Reform" (Washington, D.C.: Embassy of Mexico, Office for Press and Public Affairs, November 1991).

67. Procuraduría Agraria, *Nueva Legislación*, 13.

68. Procuraduría Agraria and Programa de Certificación de Derechos Ejidales y Titulación de Solares Urbanos, *¿Qué Es y Cómo Funciona el PROCEDE?* Crónicas del PROCEDE 8 (México, D.F.: Procuraduría Agraria and PROCEDE, 1993), 9–61.

69. See "Neoliberalismo y Campo," *Cuadernos Agrarios* 11–12 (México, D.F.: Nueva Época, January 1995).

70. The *ejido* reform allows *ejidatarios* to sell their land if the General *Ejido* Assembly so agrees in a meeting with a required quorum of 40 percent.

71. Estados Unidos Mexicanos, Poder Ejecutivo Federal, *Plan Nacional de Desarrollo 1989–1994* (México: Secretaría de Programación y Presupuesto, 1989), 69–74.

72. Procuraduría Agraria, *¿Qué Es y Cómo Funciona el PROCEDE?*, 9–13.

73. Congress of the United States, *US-Mexico Trade: Pulling Together or Pulling Apart?* (Washington, D.C.: Office of Technology Assessment, October 1992), 197.

74. Arthur Andersen & Co., *Tax and Trade Guide*: Mexico, Tax and Trade Guide Series (n.p.: 1967), 2.

75. Congress of the United States, *US-Mexico Trade*, 199.

76. David Barkin, *Distorted Development: Mexico in the World Economy* (Boulder: Westview Press, 1990), 16.

77. Ibarra Mendívil, *Propiedad Agraria*, 237.

78. Ruth Macías and José Luis Zaragoza, *El Desarrollo Agrario de México y su Marco Jurídico* (México: CNIA, 1980), 585–587, quoted in Ibarra Mendívil, Propiedad Agraria, 239–240.

79. *Caciques* are political bosses.

80. Jonathan Fox, "Political Change in Mexico's New Peasant Economy" in María Lorena Cook et at., eds., *The Politics of Economic Restructuring in Mexico: State-Society Relations and Regime Change in Mexico* (San Diego: Center for U.S.-Mexican Studies, University of California, 1994), 262.

81. Ibid., 246 and Jonathan Fox and Gustavo Gordillo, "Between State and Market: The Campesinos' Quest for Autonomy" in Wayne Cornelius et al., eds., *Mexico's Alternative Political Futures*, Monograph Series 30 (La Jolla: Center for U.S.-Mexican Studies, University of California-San Diego, 1989), 137.

82. Fox and Gordillo, "Between State and Market," 135.

83. Ibid., 136.

84. Blanco Mendoza, "Investment Implications: NAFTA," 1992.

85. Tannenbaum, *Mexico: The Struggle for Peace and Bread*, 11–12.

86. Personal communication with Daniel Coss Rangel during field research undertaken in 1994–1995. The word *huaraches* means sandals.

NOTES FOR CHAPTER THREE

1. *Sexenio* is the six-year presidential period.

2. Jorge Hardoy, *Urban Planning in Pre-Columbian America*, Planning and Cities, George R. Collins ed. (George Braziller: New York, 1968). Also see Claude Bataillon and Hélène Rivière d'Arc, *La Ciudad de México*, trans. Carlos Montemayor and Josefina Anaya (México, D.F.: SepSetentas, 1973).

3. Ibid. Also see José Luis Soberanes Reyes, *La Reforma Urbana: Una Visión de la Modernización de México* (México, D.F.: Fondo de Cultura Económica, 1993), 31.

4. Miguel León Portilla and Ángel María Garibay, *La Visión de los Vencidos: Relaciones Indígenas de la Conquista* (México, D.F.: Universidad Nacional

Autónoma de México, 1987). Also see Nick Caistor, *México City* (New York: Interlink Books, 2000), 4–13.

5. Ian Scott, *Urban and Spatial Development in Mexico* (Baltimore: World Bank, Johns Hopkins University Press, 1982), 25.

6. Bataillon and Rivière d'Arc, *La Ciudad de México,* 1973; León Portilla and Garibay, *La Visión de los Vencidos,* 1987.

7. I refer to colonial spaces as those spaces built by the colonizers within the colony. For more on colonies and colonizers see Jürgen Osterhammel, *Colonialism: A Theoretical Overview,* Trans. Shelly L. Frisch (Princeton: Markus Wiener Publishers, 1997).

8. Bataillon and Rivière d'Arc, *La Ciudad de México.* Also León Portilla, *La Visión de los Vencidos.*

9. Paul Bairoch, *Cities and Economic Development: From the Dawn of History to the Present* (Chicago: University of Chicago Press, 1988).

10. Alan Gilbert, *The Latin American City* (London and NY: Monthly Review Press, 1994). Also see Bryan R. Roberts, *The Making of Citizens: Cities of Peasants Revisited* (London and New York: Arnold, 1995), 157–187.

11. Jürgen Osterhammel, *Colonialism,* 10–11.

12. Rodolfo Puiggrós, *La España Que Conquistó al Nuevo Mundo,* (México, D.F.: Costa Amic, 1983).

13. Ibid.

14. Chapter Two lays out the history of the continuous privatization and dispossession of Indigenous lands that were enacted through different laws by the Spanish Crown.

15. The War of Independence attempted to gain local control of the economy. The local control passed from the hands of Spanish to the hands of *Criollos.* The Indians did not gain control of that economy. So, the colonial relation persisted.

16. This region is located between the states of Guanajuato and Querétaro.

17. To compare different Spanish groups see: José Tomás Falcón Gutiérrez, *Minería, Comercio y Poder: Los Criollos en el Desarrollo Económico y Político del Guanajuato de las Postrimerías del Siglo XVIII* (Guanajuato, México: La Rana, 1998). Oscar Flores, *México Minero 1796–1950: Empresarios, Trabajadores e Industria* (Monterrey, México: Editorial Font, 1994). Laura Pérez Rosales, *Minería y Sociedad en Taxco Durante el Siglo XVIII* (México, D.F.: Universidad Iberoamericana, 1996).

18. Eric W. Wolf, "El Bajío en el Siglo XVIII: Un Análisis de Integración Cultural" in David Barkin, ed., *Los Beneficiarios del Desarrollo Regional* (México, D.F.: SepSetentas 52, Edimex), 65–67.

19. Bryan R. Roberts, *Cities of Peasants* (London: Edward Arnold, 1978).

20. Michael Johns, *The City of Mexico in the Age of Díaz* (Austin: University of Texas Press, 1997), 12.

21. Ibid., 3.

22. Roberts, *The Making of Citizens,* 95–103.

23. Johns, *The City of Mexico in the Age of Díaz,* 15.

24. Scott, *Urban and Spatial Development in Mexico,* 33.

25. Ibid., 31. Also Héctor Aguilar Camín and Lorenzo Meyer, *In the Shadow of the Mexican Revolution: Contemporary Mexican History, 1910–1989* (Austin: University of Texas Press, 1993), 3.

26. Scott, *Urban and Spatial Development in Mexico*, 31.

27. Ibid., 32.

28. Johns, *The City of Mexico in the Age of Díaz*, 35–37.

29. Manuel Castells defines *vecindades* as "multifamily rental housing units, privately owned and generally located in the oldest portion of Mexico City." See Manuel Castells, *The City and the Grassroots: A Cross-Cultural Theory of Urban Social Movements* (Berkeley: University of California Press, 1983), 189.

30. Johns, *The City of Mexico in the Age of Díaz*, 53–55.

31. Ibid., 39.

32. Viviane Brachet-Marquez, *The Dynamics of Domination. State, Class, and Social Reform in Mexico 1910–1990* (Pennsylvania: University of Pittsburgh Press, 1994), 58.

33. David Cymet, *From Ejido to Metropolis, Another Path: An Evaluation on Ejido Property Rights and Informal Land Development in Mexico City*, American University Studies, Series 21, Regional Studies, Vol. 6 (New York: Peter Lang, 1992), 124.

34. *Colonias proletarias* mean literally proletarian neighborhoods or low-income neighborhoods.

35. Soberanes, *La Reforma Urbana*, 40–41.

36. Scott, *Urban and Spatial Development in Mexico*, 41.

37. Ibid., 45.

38. Cárdenas was a native of the Southcentral state of Michoacán.

39. Brachet-Marquez, *The Dynamics of Domination*, 74.

40. Cymet, *From Ejido to Metropolis*, 125–127.

41. Scott, *Urban and Spatial Development in Mexico*, 46.

42. Soberanes, *La Reforma Urbana*, 43.

43. Brachet-Marquez, *The Dynamics of Domination*, 78.

44. Aguilar Camín and Meyer, *In the Shadow of the Mexican Revolution*, 138.

45. Brachet-Marquez, *The Dynamics of Domination*, 77.

46. The import substitution industrialization consisted in the production of goods instead of importing them.

47. Julie A. Erfani, *The Paradox of the Mexican State: Rereading Sovereignty from Independence to NAFTA* (Boulder, Colorado: Lynne Rienner Publishers, 1995), 70.

48. *México a Través de los Informes Presidenciales (Política Agraria)*, quoted in Alejandro Carrillo Castro, *La Reforma Administrativa en México: Evolución de la Reforma Administrativa en México (1971–1979)* (México, D.F.: Miguel Ángel Porrúa, 1980), 62, footnote 14. Quoted in Erfani, *The Paradox of the Mexican State*, 70.

49. See Chapter Two, pages 33–34 of this book.

50. Cymet, *From Ejido to Metropolis*, 130.

51. William Simbieda, "Social Land and Urban Needs: Ejido Transformation at the Periphery," Paper presented in the Conference *The Urbanization of the Ejido: The Impact of the Reform to Article 27 Upon Real Estate Development and Land Regulation Policies* (Austin: University of Texas at Austin, photocopy, February 1994), 1–10.

52. In Mexico, "popular sector" refers to the working class population.

53. Cymet, *From Ejido to Metropolis*, 79 and 145.

54. Ibid., 169–172.

55. Although *avecindados* were allowed to acquire land in the *ejido*, they were not able to participate in the decision making of the *ejido*.

56. Alberto Rébora Togno, "Notas Sobre la Acción Gubernamental en el Combate al Precarismo," Paper presented in the Conference *The Urbanization of the Ejido: The Impact of the Reform to Article 27 Upon Real Estate Development and Land Regulation Policies* (Austin: University of Texas, photocopy, February 1994), 11.

57. Brachet-Marquez, *The Dynamics of Domination*, 98.

58. Aguilar Camín and Meyer, *In the Shadow of the Mexican Revolution*, 168.

59. Erfani, *The Paradox of the Mexican State*, 64.

60. Ibid., 60–61.

61. Kirsten Appendini and Daniel Murayama, "Desarrollo Desigual en México (1900 Y 1960)," in Barkin, *Los Beneficiarios del Desarrollo Regional*, 149.

62. Brachet-Marquez, *The Dynamics of Domination*.

63. François Perroux, "Economic Space: Theory and Applications," in John Friedmann and William Alonso, eds., *Regional Development Planning: A Reader* (Cambridge, Mass.: MIT Press, 1964), 21–36. Also see D.F. Darwent, "Growth Poles and Growth Centers in Regional Planning: A Review," in John Friedmann and Alonso William, eds., *Regional Policy: Readings in Theory and Applications* (Cambridge, Mass.: MIT Press, 1975), 538–560.

64. Soberanes, *La Reforma Urbana*, 49.

65. Ibid., 47.

66. Daniel Hiernaux, "De Frente a la Modernización: Hacia una Nueva Geografía de México," in Mario Bassols, ed., *Campo y Ciudad en una Era de Transición: Problemas, Tendencias y Desafíos* (México, D.F.: Universidad Autónoma Metropolitana-Iztapalapa, 1994), 28.

67. Ibid., 28.

68. Ibid., 27.

69. Erfani, *The Paradox of the Mexican State*, 112.

70. Ibid., 114.

71. Antonio Azuela de la Cueva, "Las Políticas de Regularización de la Ciudad de México," in Antonio Azuela and François Thomas, eds., *El Acceso de los Pobres al Suelo Urbano* (México, D.F.: Instituto de Investigaciones Sociales, Universidad Nacional Autónoma de México, 1997), 225–231.

72. Rébora Togno, "Notas Sobre la Acción Gubernamental," 17.

73. Oscar López Velarde Vega and Catalina Rodríguez Rivera, "Urbanización del Ejido: El Impacto de la Reforma del Artículo 27 Constitucional.

Planeación y Respuestas Institucionales a los Procesos de Desarrollo de Tierra Ejidal," Paper presented in the Conference *The Urbanization of the Ejido: The Impact of the Reform to Article 27 Upon Real Estate Development and Land Regulation Policies* (Austin: University of Texas, photocopy, February 1994), 6.

74. Ibid., 2.
75. Patrice Melé, "Solicitamos Se Nos Regule y No Se Nos Desaloje: Reservas Territoriales, Expropiaciones de Tierra Ejidales y Planeación Urbana," Paper presented in the Conference *The Urbanization of the Ejido: The Impact of the Reform to Article 27 Upon Real Estate Development and Land Regulation Policies* (Austin: University of Texas, photocopy, February 1994), 14–16.
76. Cymet, *From Ejido to Metropolis,* 90–94.
77. Castells, *The City and The Grassroots,* 194.
78. Ibid., 211.
79. Ibid., 212.
80. Chapter Four of this book presents a more detailed analysis of neoliberalism.
81. López Velarde Vega and Rodríguez Rivera, "Urbanización del Ejido," 4.
82. Rébora Togno, "Notas Sobre la Acción Gubernamental," 20.
83. Ibid., 23–24.
84. Aguilar Camín and Meyer, *In the Shadow of the Mexican Revolution,* 228–231.
85. Hiernaux, "De Frente a la Modernización," 32.
86. José Aranda Sánchez, "La Política Regional en México: Los Programas Estratégicos 1983–1988," in José Luis Calva, *Desarrollo Regional y Urbano: Tendencias y Alternativas,* first edition, vol. 1 (México, D.F.: Instituto de Geografía, Universidad Nacional Autónoma de México; Centro Universitario de Ciencias Sociales y Humanidades, Universidad de Guadalajara and Juan Pablos Editor, 1995), 43.
87. Aguilar Camín and Meyer, *In the Shadow of the Mexican Revolution,* 231.
88. Soberanes, *La Reforma Urbana,* 74–75.
89. Rébora Togno, "Notas Sobre la Acción Gubernamental," 27.
90. López Velarde Vega and Rodríguez Rivera, "Urbanización del Ejido," 5.
91. Rébora Togno, "Notas Sobre la Acción Gubernamental," 30.
92. Ibid., 33–34.
93. Soberanes, *La Reforma Urbana,* 117.
94. Secretaría de Desarrollo Social, SEDESOL, *Programa de 100 Ciudades 1990–2000,* (January 1993), n.p., quoted in Rébora Togno, "Notas Sobre la Acción Gubernamental," 4.
95. López Velarde Vega and Rodríguez Rivera, "Urbanización del Ejido," 20.
96. Ibid., 21.
97. Procuraduría Agraria, *Nueva Legislación Agraria,* 65.
98. Transparency in this context refers to an honest process.
99. López Velarde Vega and Rodríguez Rivera, "Urbanización del Ejido," 7.
100. Ibid.," 7–9.

101. Plats are the plans or maps of the measured lands.

102. *Comuneros* are owners of communal lands.

103. López Velarde Vega and Rodríguez Rivera, "Urbanización del Ejido," 15.

104. Appendini and Murayama, "Desarrollo Desigual en México," 127–128.

105. Nicolás López Tamayo, "La Urbanización de los Ejidos en la Ciudad de México," Paper presented in the Conference *The Urbanization of the Ejido: The Impact of the Reform to Article 27 Upon Real Estate Development and Land Regulation Policies* (Austin: University of Texas, photocopy, February 1994), 1.

NOTES FOR CHAPTER FOUR

1. Ravi Ramamurti, "Privatization and the Latin American Debt" in Robert Grosse, ed., *Private Sector Solutions to the Latin American Debt Problem* (New Brunswick and London: North-South Center, University of Miami and Transactions Publishers, 1992), 164.

2. Hermann Sautter, ed. *Economic Reforms in Latin America* (Frankfurt and Main: Symposium held in November 1992 at the Georg-August-Universität Göttingen, Vervuert Verlag, 1993), 10.

3. Arturo Escobar, *Encountering Development: The Making and Unmaking of the Third World* (Princeton, NJ: Princeton University Press, 1995), 80.

4. On the Mexican debt and crisis see Carlos Bazdrech, Nisso Bucay, Soledad Loaeza and Nora Lustig, *México: Auge, Crisis y Ajuste,* Macroeconomía y Deuda Externa, 1982–1989, vol. 2 (México, D.F.: Fondo de Cultura Económica, 1992). Vladimiro Brailowsky, "Las Implicaciones Macro-económicas de Pagar: La Política Económica Antes de la 'Crisis' de la Deuda en México" in Bazdrech et al., idem. Also see Grosse, *Private Sector Solutions.* Sautter, *Economic Reforms in Latin America.* Escobar, *Encountering Development.* Gerardo Otero, *Neoliberalism Revisited: Economic Restructuring and Mexico's Political Future* (Simon Fraser University and Westview Press, 1996).

5. Manuel Lasaga, "The Road Back to the Marketplace: How Banks Have Managed the LDC Debt-Restructuring Process" in Grosse, *Private Sector Solutions,* 16.

6. For a more ample discussion on the Northern and Southern agendas, see Chapter Two of this book.

7. Stephany Griffit-Jones and Osvaldo Sunkel, *Debt and Development Crisis in Latin America: The End of an Illusion* (Oxford, Clarendon: Oxford University Press, 1989).

8. Clark W. Reynolds, "¿Una Generación Perdida? ¿Por Qué el Desarrollo Latinoamericano Depende del Crecimiento?" in Bazdrech et al, *México: Auge, Crisis, y Ajuste,* 246.

9. See Escobar, *Encountering Development.* Also Jonathan Crush, *Power of Development* (London and New York: Routledge, 1995).

10. José Ángel Gurría T., "La Política de Deuda Externa de México, 1982–1990" in Brazdrech et al., *México: Auge, Crisis, y Ajuste,* 294–295.

11. Otero, *Neoliberalism Revisited*, 7.
12. Lasaga, "The Road Back to the Marketplace," 13–15.
13. Ibid., 21.
14. Ibid., 22.
15. Vladimir Ilich Lenin, *Imperialism: The Highest Stage of Capitalism* (New York: International Publishers, 1939), 53.
16. Gross, *Private Sector Solutions*, 7.
17. Ibid., 9.
18. "1989 Regulations of the Law to Promote Mexican Investment and to Regulate Foreign Investment" English translation in *Doing Business in Mexico* (New York: Price Waterhouse, 1992).
19. Otero, *Neoliberalism Revisited*, 7.
20. Miguel D. Ramírez, "The Political Economy of Privatization in Mexico," Occasional Paper No. 1, presented at the conference *Mexico Beyond NAFTA* (Amherst: University of Massachusetts, Latin American Consortium of New England, n.d.), 27–28.
21. Elvira Concheiro Bójorquez, *El Gran Acuerdo*, (México, D.F.: Instituto de Investigaciones Económicas, Universidad Nacional Autónoma de México (UNAM), and ERA, 1996), 12.
22. Bazdrech et al., *México: Auge, Crisis y Ajuste*, 9.
23. Alain De Janvry, Gustavo Gordillo and Elisabeth Sadoulet, *Mexico's Second Agrarian Reform. Household and Community Responses* (San Diego: U.S. Mexican Studies, University of California, 1997), 13–22.
24. Concheiro Bójorquez, *El Gran Acuerdo*, 12–36.
25. Ibid., 37.
26. Pedro Aspe Armella, *El Camino Mexicano de la Transformación Económica*, first repr. (México, D.F.: Fondo de Cultura Económica, 1993), 162.
27. John Vickers and George Yarrow, *Privatization: An Economic Analysis* (Mass.: MIT Press, Series on the Regulation of Economic Activity #18, 1988), cited by Ramamurti, "Privatization and the Latin American Debt," 168.
28. Ramamurti, "Privatization and the Latin American Debt," 170.
29. Ibid., 153.
30. Sautter, *Economic Reforms in Latin America*, 10.
31. Ibid., 23.
32. Sidney Weintraub, ed., *Integrating the Americas: Shaping Future Trade Policy* (New Brunswick: Transaction Publishers and North/South Center, University of Miami, 1994), xii.
33. Ross Schneider, Ben, "The Politics of Privatization in Brazil and Mexico: Variations on a Statist Theme," Conference Paper No. 23, (New York: A National Resource Center for Latin American and Caribbean Studies, The Columbia University/New York University Consortium, 1990), 27–30.
34. The de la Madrid administration went from 1982 through 1988. It was during this period that the debt crisis was declared.
35. Ross Schneider, "The Politics of Privatization," 31.

36. Luis Echeverría was president for the 1970–1976 *sexenio.*
37. López Portillo's term was from 1976 to 1982.
38. Bazdresch et al., *México: Auge, Crisis y Ajuste*, 9.
39. Ross Schneider, "The Politics of Privatization," 32.
40. The *Partido Revolucionario Institucional* or Revolutionary Institutional Party (PRI) was founded by ex-president Plutarco Elías Calles in 1929 under the name of Revolutionary National Party (PNR). In 1946 the PNR became the PRI, the "official" party and the one that uninterruptedly won all elections from 1929 through 1988. See Kevin J. Middlebrook, *The Paradox of the Mexican Revolution,* (Baltimore and London: The Johns Hopkins University Press, 1995), 27.
41. Ross Schneider, "The Politics of Privatization," 32.
42. Ramírez, "The Political Economy of Privatization," 12–14.
43. Ibid., 14–20.
44. Ibid., 16.
45. Salinas de Gortari's presidential period went from 1988 to 1994.
46. Talk with IMF Economist at Cornell University, 1997.
47. Ramírez, "The Political Economy of Privatization," 16.
48. Ibid., 20.
49. Ibid., 11–12.
50. Ibid., 20.
51. Sautter, *Economic Reforms in Latin America*, 16.
52. Marilyn Gates, "The Debt Crisis and Economic Restructuring: Prospects for Mexican Agriculture," in Otero, *Neoliberalism Revisited*, 51, 57 and 59.
53. Estados Unidos Mexicanos. Poder Ejecutivo Federal, *Plan Nacional de Desarrollo, 1983–1989* (México, D.F.: Secretaría de Programación y Presupuesto, 1983).
54. See Rammamurti, "Privatization and the Latin American Debt"; Grosse, *Private Sector Solutions*; Lasaga, "The Road Back to the Marketplace"; Concheiro, *El Gran Acuerdo*; and Ross Schneider, "The Politics of Privatization".
55. Julie A. Erfani, *The Paradox of the Mexican State*, 180.
56. In this case, *mestizaje* refers to the mixing of the Indian and Spanish cultures.
57. DeWalt and Rees with Murphy, *The End of Agrarian Reform*, 2.
58. Ramírez, "The Political Economy of Privatization," 30.
59. Sabino Cassese, "Public Enterprises After Privatization," in Alvaro Antonio Zini, *The Market and the State in Economic Development in the 1990s* (North Holland, B.V.: Elsevier Science Publisher, 1992), 70–71.
60. Ramírez, "The Political Economy of Privatization," 22–23.
61. Aspe Armella, *El Camino Mexicano*, 172.
62. Ramírez, "The Political Economy of Privatization," 34–35.
63. Erfani, *The Paradox of the Mexican State*, 33–58 and 180.
64. Asociación de Banqueros de México, "Informe del Consejo Directivo a la Convención Anual," (México, 1973), 5, quoted by Concheiro Bójorquez, *El Gran Acuerdo*, 181. [My translation.]
65. Concheiro Bójorquez, *El Gran Acuerdo*, 56–57.
66. *Porfiriato* is the era of Porfirio Díaz.

67. Coparmex, *Franco Diálogo Entre Gobierno y Empresarios,* (México, 1971), 14–15, quoted in Concheiro Bójorquez, *El Gran Acuerdo,* 56–57. [My translation.]
68. Grosse, *Private Sector Solutions,* 183.
69. Concheiro Bójorquez, *El Gran Acuerdo,* 60.
70. Ibid., 62.
71. Otero, *Neoliberalism Revisited,* 14.
72. Aspe Armella, *El Camino Mexicano,* 175.
73. Brailowsky, "Las Implicaciones Macroeconómicas de Pagar," 123.
74. Ibid., 125.
75. Ibid., 126–127.
76. Ross Schneider, "The Politics of Privatization," 24.
77. Weintraub, *Integrating the Americas,* iii.
78. Otero, *Neoliberalism Revisited,* 4.
79. Lasaga, "The Road Back to the Marketplace," 31.
80. Robert Gilpin, *The Political Economy of International Relations,* (Princeton, NJ: Princeton University Press, 1988), 177.
81. Gert Preuße, "Mexico and the North American Free Trade Association (NAFTA)," in Sautter, *Economic Reforms in Latin America,* 132.
82. Escobar, *Encountering Development,* 3–4.
83. Manfred Bienfeld, "Structural Adjustment: Where Do We Go From Here?," Talk delivered at Cornell University (November 4, 1993).
84. Preuße, "Mexico and the North American Free Trade," 145–149.
85. Ibid., 139.
86. Peter Nunnenkamp, "Economic Policies and Attractiveness for Foreign Capital: The Experience of Highly Indebted Latin American Countries," in Sautter, *Economic Reforms in Latin America,* 91.
87. Otero, *Neoliberalism Revisited,* 1.
88. David Ricardo, "On Foreign Trade" in *The Principles of Political Economy and Taxation,* First Published in 1821, Third Edition, Amherst and New York: Prometheus Books, 1996. Cited by Gilpin, *The Political Economy of International Relations,* 173–174.
89. Gilpin, *The Political Economy of International Relations,* 173.
90. Paul Krugman, *Strategic Trade Policy and The New International Economics* (Cambridge: MIT Press, 1986), 8. Quoted by Gilpin, *The Political Economy of International Relations,* p. 223.
91. Specialization can also be by product.
92. Paul Krugman, "Economies of Scale, Imperfect Competition, and Trade: An Exposition," Unpublished paper, 1981, 2. Cited by Gilpin, *The Political Economy of International Relations,* 176–177.
93. Howard M. Watchel, *The Money Mandarins: The Making of a Supranational Economic Order.* New York: Sharpe, Inc., 1990), 4–8 and 183.
94. Herb Addo, *Imperialism: The Permanent Stage of Capitalism,* World-System Critique of Eurocentric Conceptions in Political Economy, vol. 1 (Tokyo, Japan: The United Nations University, 1986), 98.

95. Lenin, *Imperialism*, 89.
96. Malcolm Waters, *Globalization* (London and New York: Routledge, 1995).
97. These characteristics have been partially developed based on class discussions at Cornell University during the period 1993–1995. Especially during the course with Lourdes Benería (Development and Change in the Global Economy, 1993) and the course with Phil McMichael (The Sociology of the World Economy, 1993). Another part was conceptualized during the discussions of the Conference on Globalization organized by Professor McMichael. This conference took place at Cornell University in the fall of 1998.
98. Addo, *Imperialism: The Permanent Stage of Capitalism*, Op. Cit.; Fernand Braudel, *Afterthoughts on Material Life*, (Baltimore, MD: Johns Hopkins University Press, 1979); and Jane M. Jacobs, *Edge of Empire: Postcolonialism and the City* (London and New York: Routledge, 1996).
99. Addo, *Imperialism: The Permanent Stage of Capitalism*, 10.
100. Braudel, *Afterthoughts on Material Life*, 111.
101. Ibid., 111–112.
102. Addo, *Imperialism: The Permanent Stage of Capitalism*, 14.
103. Braudel, *Afterthoughts on Material Life*, 113–114.
104. Addo, *Imperialism: The Permanent Stage of Capitalism*, 10.
105. José Maria Fanelly, Roberto Frenkel, and Guillermo Rozenwurcel, "Growth and Structural Reform in Latin America: Where We Stand" in Zini, *The Market and The State*, 245–246.
106. Jacobs, *Edge of Empire*, 16.
107. Edward W. Said, *Orientalism* (New York: Vintage Books, 1979), 331.
108. Jacobs, *Edge of Empire*, 35.

NOTES FOR CHAPTER FIVE

1. Procuraduría Agraria, *Ejido San Luis, El Pionero del Procede*, photocopied material provided by the Procuraduría Agraria (Hermosillo, Sonora: México, August, 1994).
2. Assembly line factories.
3. Barry, *Mexico: A Country Guide*, 142.
4. These export-based border assembly plants are owned mainly by U.S. and Asian capital. Most of them assemble parts for the electronics industry and employ a considerable percentage of female labor force. For more on *maquiladoras* see Devon Peña, *The Terror of the Machine: Technology, Work, Gender, & Ecology on the U.S.-Mexico Border* (Austin: Center for Mexican American Studies, the University of Texas, 1997).
5. David E. Lorey, *The U.S.-Mexican Border in the Twentieth Century* (Wilmington: Scholarly Resources, 1999), 89–91.
6. To review the definition of "restricted areas", see the section "Who Could Own Land and Where" in Chapter Two of this research.
7. Lorey, *The U.S.-Mexican Border in the Twentieth Century*, 106.
8. Ibid., 107.

9. Procuraduría Agraria, *San Luis Río Colorado, Sonora*, Crónicas de Procede #3 (México: Imprenta Juventud, Procuraduría Agraria, Mayo 1994).

10. Miguel Ángel Vázquez Ruiz ed., *Sonora hacia el 2000: Tendencias y Desafíos*. (Hermosillo, Sonora: Ed. Sino, 1993).

11. David Myhre, "Appropriate Agricultural Credit: A Missing Piece of Agrarian Reform in Mexico" in Randall, *Reforming Mexico's Agrarian Reform*, 118.

12. Daniel Covarrubias Patiño, "An Opinion Survey in the Countryside–1994" in Randall, *Reforming Mexico's Agrarian Reform*, 113.

13. Myhre, "Appropriate Agricultural Credit," 118.

14. Luin Goldring, "The Changing Configuration of Property Rights Under Ejido Reform" in Randall, *Reforming Mexico's Agrarian Reform*, 276.

15. Procuraduría Agraria, *Ejido San Luis. El Pionero del Procede*, 11.

16. Instituto Nacional de Estadística, Geografía e Informática (INEGI), *San Luis Río Colorado, Estado de Sonora. Cuaderno Estadístico Municipal, Edición 1994* (Aguascalientes, México: INEGI, 1995), 63–70.

17. INEGI, *San Luis Río Colorado, Cuaderno Estadístico Municipal*, 57.

18. Procuraduría Agraria, *San Luis Río Colorado*, 11.

19. Ibid., 11.

20. Procuraduría Agraria, *Ejido San Luis, El Pionero del Procede*.

21. For more on "regularization", see Chapter Three of this research.

22. Procuraduría Agraria, *San Luis Río Colorado, Sonora*. 18–19 and 22–23.

23. Ibid., 27.

24. Personal interview with Enrique Orozco Oceguera, November 12, 1994.

25. Gustavo Garza, *Desconcentración, Tecnología y Localización Industrial en México: Los Parques y Ciudades Industriales, 1953–1988*. México, D.F.: El Colegio de México, 1992), 147.

26. All the references to Orozco Oceguera come from my personal interview with him.

27. Note the paradoxical name of the enterprise, taking into consideration that the *ejido* is not private property and that by creating such an enterprise the *ejido* simply stopped being an *ejido*.

28. Procuraduría Agraria, *San Luis Río Colorado, Sonora*, 20–21.

29. At that time one hundred new pesos was the equivalent of $33 dollars. Currently (year 2000) is the equivalent of $12 dollars.

30. *Ley General de Sociedades Mercantiles; Ley General de Sociedades Cooperativas* (México, D.F.: Editorial Pac, 1995).

31. Procuraduría Agraria, *San Luis Río Colorado, Sonora*, 28–29.

32. Personal interview with Alberto Ávila Soto, *ejidatario* of SLRC, November 4, 1994.

33. Information obtained during informal interviews with *ejidatarios*.

34. Personal interview with Felipa Haro, *ejidataria* of SLRC, November 10, 1994.

35. *Comisión de Regularización y Tenencia de la Tierra* or Regularization and Land Tenure Commission was the agency in charge of regularizing expropriated land for urban uses. In SLRC, CORETT and not the Procuraduría Agraria had the responsibility of issuing land certificates. For more on CORETT see Chapter Three of this research.

36. Procuraduría Agraria, *San Luis Río Colorado, Sonora,* 41.
37. Personal communication with Daniel Coss Rangel during field research undertaken in 1994–1995. The word *huaraches* means sandals.
38. Data from interview undertaken in the *ejido*, November 1994.
39. This is the datum obtained in the interviews with the women of the *ejido*. This datum, however, does not correspond to the 1990 Census Data that specifies as 2.5 the average of kids per women above 12 years old.

NOTES FOR CHAPTER SIX

1. Since residents of the *ejido* used La Poza more than La Zanja, I also decided to use La Poza to refer to the *ejido*.
2. José Reveles, "Punta Diamante, Contraste de Riqueza y Miseria," *El Financiero*, Saturday, 12 November 1994, 1, 22–23.
3. Instituto Nacional de Estadística, Geografía e Informática (INEGI), *Acapulco de Juárez, Estado de Guerrero: Cuaderno Municipal, Edición 1993,* (Aguascalientes, México: INEGI, 1994.
4. Humberto Aburto Parra, "Guión Metodológico para el Proyecto de Desarrollo Rural Integral," draft report provided by the author in November, 1994, 5.
5. Interview with the local practitioner in La Clínica de Solidaridad "La Poza", December, 1994.
6. For more on the African-Mexican population of Guerrero, see Gonzalo Aguirre Beltrán, *Cuijila: Esbozo Etnográfico de un Pueblo Negro* (México, D.F.: Fondo de Cultura Económica, 1985) and Gonzalo Aguirre Beltrán, *La Población Negra de México: Estudio Etnográfico* (México, D.F.: Fondo de Cultura Económica, 1984).
7. In Spanish they are called Acapulco *Tradicional*, Acapulco *Dorado*, and Acapulco *Diamante*.
8. Reveles, "Punta Diamante, Contraste de Riqueza y Miseria," 1, 22–23.
9. Francisco A. Gomezjara, *Bonapartismo y Lucha Campesina en la Costa Grande de Guerrero* (México, D.F.: Editorial Posada, 1979).
10. Ibid., 188.
11. Ibid., 188–192.
12. Ibid., 192–195.
13. *Palpitaciones Porteñas*, Acapulco, Guerrero, México, November 1949, quoted by Gomezjara, *Bonapartismo y Lucha Campesina en la Costa Grande de Guerrero*, 191.
14. Gomezjara, *Bonapartismo y Lucha Campesina en la Costa Grande de Guerrero*, 194.
15. Interview with *ejidatarios* of La Poza.
16. Reveles, "Punta Diamante, Contraste de Riqueza y Miseria," 1, 22–23.
17. *Promotora Turística de Guerrero (PROTUR)*, photocopied material provided by PROTUR, n/d, n/p.
18. For more on Territorial Reserves see Chapter Three of this research.
19. PROTUR, *Acapulco Diamante: El Acapulco del Siglo XXI*, promotional material provided by PROTUR, n/d, n/p.

20. Information provided by the *Comisariado Ejidal* of La Poza in December of 1994.
21. The *Carretera del Sol* or Highway of the Sun is one kilometer from the *ejido* La Poza.
22. Interview with Benjamín Sandoval Melo, PRD local representative, Acapulco, Guerrero, December, 1994. The PRD is the *Partido de la Revolución Democrática* or Party of the Democratic Revolution.
23. Documents read at the home of the *Comisariado Ejidal* and not available for photocopying.
24. In December 1994, the exchange rate was 3,000 pesos per dollar. Thus, she was selling her land for about $333.
25. Mexican shawl.
26. The words for privatization and deprivation are very similar in Spanish: *privatización* and *privación*.
27. Articles from *El Sur*, Acapulco, Guerrero, April-May 1994.
28. Interview with a sales representative of the Vidafel development.
29. *Narcos* is the short name for the Spanish word *narcotraficante*, which means drug lords.
30. INEGI, *Acapulco de Juárez, Estado de Guerrero: Cuaderno Municipal, Edición 1993*, 44.
31. Interview with the *Comisariado Ejidal* of La Poza, December 1994.

NOTES FOR CHAPTER SEVEN

1. *Banco Nacional de Crédito Rural* or Rural Credit National Bank.
2. Anonymous report provided by the *ejido* authority, n/d, 3.
3. Leticia Reyna Aoyama, ed., *Economía Contra Sociedad: El Itsmo de Tehuantepec 1907–1986*, (México: Nueva Imagen, 1994.)
4. Information taken from interviews with residents of the Ixtaltepec *ejido* undertaken in September 1994.
5. Instituto Nacional de Estadística, Geografía e Informática (INEGI), *Juchitán de Zaragoza, Estado de Oaxaca, Cuaderno Estadístico Municipal, Edición 1995* (Aguascalientes, México: INEGI, 1995), 21. Also Anonymous report, 8.
6. INEGI, *Juchitán de Zaragoza, Estado de Oaxaca, Cuaderno Estadístico Municipal, Edición 1995*, 1–127.
7. Ludka de Gortari, "Comunidad Como Forma de Tenencia de la Tierra" in *Revista de la Procuraduría Agraria No. 8*, July-September (México: Procuraduría Agraria, 1997), 7.
8. Héctor Manuel Sánchez López, *Estudio Monográfico de la Comunidad de Asunción Ixtaltepec, Juchitán, Oaxaca*, photocopy, March 1975, 10.
9. Ibid., 10 and 27.
10. For more on the action taken by SLRC, see Chapter Five of this research.
11. Corn mills.
12. Cooked beverage prepared with ground corn, water, and cinnamon.

13. *Nixtamal* is the dough which results from grinding corn. It is used to prepare tortillas and atole, among other meals.
14. From the Isthmus region.
15. *Ella* is a pop magazine targeting women.
16. Reyna Aoyama, *Economía Contra Sociedad: El Itsmo de Tehuantepec 1907–1986.*
17. Initially, the Transisthmian railroad was used to transport merchandise from the Pacific to the Atlantic coasts. The construction of the Panama Canal in 1914 provoked a decline in the region as users of the railroad shifted to the Canal for transporting their goods. For more on this history, see Reyna Aoyama.
18. Ian Scott, *Urban and Spatial Development in Mexico. 67.*
19. Taken from the structured interviews undertaken in the *ejido* in 1994.

NOTES FOR CHAPTER EIGHT

1. *Ejido* San Luis's Contractors and Real Estate Development.
2. Articles from *El Sur*, April-May 1994.
3. Interview with Benjamín Sandoval Mello, Acapulco, Guerrero, December 1994.
4. Promotora Turística de Guerrero (PROTUR). Photocopied material provided by PROTUR. Also see *Acapulco Diamante: El Acapulco del Siglo XXI.* Promotional material provided by PROTUR.
5. Victor Ruiz Arrazola, "El Senado Pidió Preferir el Capital Mexicano al Transnacional: Ruiz Sacristán," in *La Jornada*, July 23, 1996.
6. Procuraduría Agraria, *Informe Anual de Actividades de la Institución* (México, D.F.: Procuraduría Agraria, 1998), 2.
7. Guillermo R. Zepeda Lecuona, "Cuatro Años de Procede: Avances y Desafíos en la Definición de Derechos Agrarios en México," *Revista de la Procuraduría Agraria* No. 9, October 1997–April 1998, 1–8.
8. Ibid., 4. These are not the words used in the referred text; rather I formulate them based on their description of methods.
9. Emma E. Aguado Herrera and Francisco Hérnandez y Puente, "Tierra Social y Desarrollo Urbano: Experiencias y Posibilidades," in *Revista de la Procuraduría Agraria*, No. 8 (México, D.F.: Julio-Septiembre, 1997), 17. Found at http://www.pa.gob.mx/publica/pa070807.htm
10. For more on the 100 Cities Program, see Chapter Three of this research.
11. Ibid., 4.
12. Ibid., 4.
13. Klaus Deininger and Hans Binswanger, *The Evolution of the World Bank's Land Policy*, (Washington, DC: World Bank, 1998), 2.
14. LRPP stands for Land Reform Policy Paper, which is the 1975 World Bank document that guided the land policy implemented as part of the structural adjustment programs.
15. Deininger and Binswanger, *The Evolution of the World Bank's Land Policy*, 2.

16. These are the Reforms that, according to supranational financial institutions, are currently needed to alleviate the socio-economic impacts of the economic restructuring policies applied in the early and mid-1990s.

17. Kevin Davis and Michael J. Trebilcock, "What Role Do Legal Institutions Play in Development?," Draft prepared for the International Monetary Fund's Conference on Second Generation Reforms, November 8–9, 1999, (np: October 20, 1999), 42.

18. David A. Atwood, "Land Registration in Africa: The Impact of Agricultural Production," in *World Development* 18, 5, 1990 659, 663, quoted by Davis and Trebilcock, "What Role Do Legal Institutions Play in Development?," 43.

19. Jean-Phillipe Platteau, *Reforming Land Rights in Sub-Saharan Africa: Issues of Efficiency and Equity* (UNRISD: United Nations Research Institute for Social Development, Discussion Paper No. 60, March 1995), 17–19, quoted by Davis and Trebilcock, "What Role Do Legal Institutions Play in Development?," 43.

20. For more on immunity, see the "Letters of Intent" signed between borrower countries and the IMF available in the IMF web site: http//www.imf.org.

21. Ibid., 43.

22. Aguado Herrera and Hernández y Puente, "Tierra Social y Desarrollo Urbano," 11.

23. Catherine Andre and Jean-Philippe Platteau, *Land Relations Under Unbearable Stress: Rwanda Caught in the Malthusian Trap*, (Namu, Belgium: Centre de Recherche en Economie du Development [CRED], 1996) quoted by Deininger and Binswanger, *The Evolution of the World Bank's Land Policy*, 3.

24. Deininger and Binswanger, *The Evolution of the World Bank's Land Policy*, 5.

25. Ibid., 12.

26. Secretaría de la Reforma Agraria, *Comunicado de Prensa* No. SRA/006, Unidad de Comunicación Social, Boletín de Prensa, January 29, 2001.

27. Secretaría de la Reforma Agraria, *Comunicado de Prensa* No. SRA/007, Unidad de Comunicación Social, Boletín de Prensa, January 31, 2001.

28. Ibid.

29. Secretaría de Desarrollo Social (SEDESOL), Colegio de Arquitectos and Instituto de Investigaciones Económicas, *México 2020. Un Enfoque Territorial de Desarrollo; Vertiente Urbana. Síntesis Ejecutiva* (México, D.F.: SEDESOL, 2000), 8–9.

30. Ibid., 42–43.

31. Ibid., 42–43.

32. Ibid., 67.

33. Ibid, 67.

34. Ibid, 68.

35. Hermann Bellinghausen, "¿Por Qué Marchan los Comandantes del EZLN?," *La Jornada in Internet*, February 1, 2001, htpp://www.jornada.unam.mx/2001/feb01/010217/ezln/html, 8.

36. Ibid, 6.

37. Ibid, 10.

Bibliography

"1989 Regulations of the Law to Promote Mexican Investment and to Regulate Foreign Investment." English translation in *Doing Business in Mexico*, New York: Price Waterhouse, 1992.

"Neoliberalismo y Campo." *Cuadernos Agrarios* 11–12. México, D.F.: Nueva Época, January 1995.

"PROCAMPO." In *Programas de Apoyo*. Web page of the Secretaría de Agricultura, Ganadería, Desarrollo Rural, Pesca y Alimentación [Secretary of Agriculture, Cattle Raising, Rural Development, Fishing and Food). htpp://www.sagarrpa.gob.mx/sagar2.htm

Aburto Parra, Humberto. "Guión Metodológico para el Proyecto de Desarrollo Rural Integral," draft report provided by the author in November, 1994

Acosta Romero, Miguel and Genaro David Góngora Pimentel. *Constitución Política de los Estados Unidos Mexicanos*. México: Editorial Porrúa, 1984.

Addo, Herb. *Imperialism: The Permanent Stage of Capitalism*. World System Critique of Eurocentric Conceptions in Political Economy, vol. 1. Tokyo, Japan: The United Nations University, 1986.

Aguado Herrera, Emma E. and Francisco Hérnandez y Puente. "Tierra Social y Desarrollo Urbano: Experiencias y Posibilidades." In *Revista de la Procuraduría Agraria*, No. 8 (México, D.F.: Julio-Septiembre, 1997), 17. Also found at http://www.pa.gob.mx/publica/pa070807.htm

Aguilar Camín, Héctor and Lorenzo Meyer. *In the Shadow of the Mexican Revolution: Contemporary Mexican History, 1910–1989*. Austin: University of Texas Press, 1993.

Aguirre Beltrán, Gonzalo. *Cuijila: Esbozo Etnográfico de un Pueblo Negro*. México, D.F.: Fondo de Cultura Económica, 1985.

———. *La Población Negra de México: Estudio Etnográfico*. México, D.F.: Fondo de Cultura Económica, 1984.

Andersen, Arthur & Co. *Tax and Trade Guide: Mexico*. Tax and Trade Guide Series. n.p.: 1967.

Andre, Catherine and Jean-Philippe Platteau. *Land Relations Under Unbearable Stress: Rwand Caught in the Malthusian Trap*. Namu, Belgium: Centre de Recherche en Economie du Development [CRED], 1996. Quoted by Deininger, Klaus and Hans Binswanger. *The Evolution of the World Bank's Land Policy*, Washington, D.C.: World Bank, 1998.

Anonymous report on Ixtaltepec provided by the *ejido* authority, n/d.

Appendini, Kirsten and Daniel Murayama. "Desarrollo Desigual en México (1900 Y 1960)." In David Barkin, ed. *Los Beneficiarios del Desarrollo Regional*. México, D.F.: SepSetentas 52, Edimex, 63–95.

Aranda Sánchez, José. "La Política Regional en México: Los Programas Estratégicos 1983–1988." In José Luis Calva, ed. *Desarrollo Regional y Urbano. Tendencias y Alternativas*. First edition, vol. 1. México, D.F.: Instituto de Geografía, Universidad Nacional Autónoma de México; Centro Universitario de Ciencias Sociales y Humanidades, Universidad de Guadalajara and Juan Pablos Editor, 1995.

Aranda, Josefina, Carlota Botey and Rosario Robles. *Tiempo de Crisis, Tiempo de Mujeres*. Oaxaca, México: Universidad Autónoma Benito Juárez de Oaxaca, Centro de Estudios de la Cuestión Agraria Mexicana, 2000.

Asociación de Banqueros de México. "Informe del Consejo Directivo a la Convención Anual." México, 1973. Quoted in Elvira Concheiro Bójorquez. *El Gran Acuerdo*. México, D.F.: Instituto de Investigaciones Económicas, Universidad Nacional Autónoma de México (UNAM), and ERA, 1996.

Aspe Armella, Pedro. *El Camino Mexicano de la Transformación Económica*. First Repr. México, D.F.: Fondo de Cultura Económica, 1993.

Atwood, David A. "Land Registration in Africa: The Impact of Agricultural Production." In *World Development* 18, 5, 1990 659, 663, quoted by Davis, Kevin and Michael J. Trebilcock. "What Role Do Legal Institutions Play in Development?" Draft prepared for the International Monetary Fund's Conference on Second Generation Reforms, November 8–9, 1999. Np: October 20, 1999.

Azuela de la Cueva, Antonio. "Las Políticas de Regularización de la Ciudad de México." In Antonio Azuela and François Thomas, eds., *El Acceso de los Pobres al Suelo Urbano*. México, D.F.: Instituto de Investigaciones Sociales, Universidad Nacional Autónoma de México, 1997.

Azuela, Antonio and François Thomas, eds., *El Acceso de los Pobres al Suelo Urbano*. México, D.F.: Instituto de Investigaciones Sociales, Universidad Nacional Autónoma de México, 1997.

Bairoch, Paul. *Cities and Economic Development: From the Dawn of History to the Present*. Chicago: University of Chicago Press, 1988.

Barkin, David, ed. *Los Beneficiarios del Desarrollo Regional.* México, D.F.: SepSetentas 52, Edimex, 63–95.

Barkin, David. *Distorted Development: Mexico in the World Economy.* Boulder: Westview Press, 1990.

Barry, Tom, ed. *Mexico: A Country Guide.* New Mexico: The Inter-Hemispheric Education Resource Center, 1992.

Bartra, Roger. *Estructura Agraria y Clases Sociales en México.* México, D.F.: Ediciones ERA and Instituto de Investigaciones Sociales de la Universidad Nacional Autónoma de México, 1974, repr., 1991.

Bassols, Mario, ed. *Campo y Ciudad en una Era de Transición. Problemas, Tendencias y Desafíos.* México, D.F.: Universidad Autónoma Metropolitana-Iztapalapa, 1994.

Bataillon, Claude and Hélène Rivière d'Arc. *La Ciudad de México.* Trans. Carlos Montemayor and Josefina Anaya. México, D.F.: SepSetentas, 1973.

Bazdrech, Carlos, Nisso Bucay, Soledad Loaeza and Nora Lustig. *México: Auge, Crisis y Ajuste.* Macroeconomía y Deuda Externa, 1982–1989, vol. 2. México, D.F.: Fondo de Cultura Económica, 1992.

Bellinghausen, Hermann. "Por Qué Marchan los Comandantes del EZLN?" *La Jornada in Internet,* February 1, 2001. http://www.jornada.unam.mx/2001/feb01/010217/ezln.html

Bienfeld, Manfred. "Structural Adjustment: Where Do We Go From Here?" Talk delivered at Cornell University, November 4, 1993.

Blanco Mendoza, Herminio. "Investment Implications: NAFTA." In *Investing in Mexico.* Report No. 999. New York: The Conference Board, 1992.

Brachet-Marquez, Viviane. *The Dynamics of Domination. State, Class, and Social Reform in Mexico 1910–1990.* Pennsylvania: University of Pittsburgh Press, 1994.

Brailowsky, Vladimiro. "Las Implicaciones Macroeconómicas de Pagar: La Política Económica Antes de la 'Crisis' de la Deuda en México, 1982–1988." In Carlos Bazdrech, Nisso Bucay, Soledad Loaeza and Nora Lustig. *México: Auge, Crisis y Ajuste.* Macroeconomía y Deuda Externa, 1982–1989, vol. 2. México, D.F.: Fondo de Cultura Económica, 1992.

Braudel, Fernand. *Afterthoughts on Material Life.* Baltimore, MD: Johns Hopkins University Press, 1979.

Calva, José Luis, ed. *Desarrollo Regional y Urbano. Tendencias y Alternativas.* First edition, vol. 1. México, D.F.: Instituto de Geografía, Universidad Nacional Autónoma de México; Centro Universitario de Ciencias Sociales y Humanidades, Universidad de Guadalajara and Juan Pablos Editor, 1995.

Carrillo Castro, Alejandro. *La Reforma Administrativa en México: Evolución de la Reforma Administrativa en México (1971–1979).* México, D.F.: Miguel Ángel Porrúa, 1980.

Cassese, Sabino. "Public Enterprises After Privatization" in Alvaro Antonio Zini. *The Market and the State in Economic Development in the 1990s*. North Holland, B.V.: Elsevier Science Publisher, 1992.

Castells, Manuel. *The City and The Grassroots: A Cross-Cultural Theory of Urban Social Movements*. Berkeley: University of California Press, 1983.

Castoir, Nick. *Mexico City*. Foreword by Elena Poniatowska. New York: Interlink Books, 2000.

Concheiro Bójorquez, Elvira. *El Gran Acuerdo*. México, D.F.: Instituto de Investigaciones Económicas, Universidad Nacional Autónoma de México (UNAM), and ERA, 1996.

Congress of the United States. *US-Mexico Trade: Pulling Together or Pulling Apart?* Washington, D.C.: Congress of the United States. Office of Technology Assessment, October 1992.

Constitución Política de los Estados Unidos Mexicanos: Comentada. Serie: A Fuentes.-b) Textos y Estudios Legislativos no. 59. México: Colección Popular Ciudad de México, Serie Textos Jurídicos, Instituto de Investigaciones Jurídicas de la Universidad Nacional Autónoma de México, Procuraduría General de Justicia del Distrito Federal, 1992.

Cook, María Lorena, Kevin J. Middlebrook, and Juan Molinar Horcasitas, eds. *The Politics of Economic Restructuring in Mexico: State-Society Relations and Regime Change in Mexico*. San Diego: Center for U.S.-Mexican Studies, University of California, 1994.

Coparmex. *Franco Diálogo Entre Gobierno y Empresarios*. México, 1971. Quoted in Concheiro Bójorquez, Elvira. *El Gran Acuerdo*. México, D.F.: Instituto de Investigaciones Económicas, Universidad Nacional Autónoma de México (UNAM), and ERA, 1996.

Cornelius, Wayne and David Myhre, eds. *The Transformation of Rural Mexico: Reforming the Ejido Sector*. San Diego: Center for U.S.-Mexican Studies, University of California, 1998.

Cornelius, Wayne, Judith Gentleman and Peter H. Smith, eds. *Mexico's Alternative Political Futures*. Monograph Series 30. La Jolla: Center for U.S.-Mexican Studies, University of California-San Diego, 1989.

Covarrubias Patiño, Daniel. "An Opinion Survey in the Countryside–1994." In Laura Randall ed. *Reforming Mexico's Agrarian Reform*, Armonk, New York: M. E. Sharpe, 1996.

Cox, Robert. "Gramsci, Hegemony and International Relations: An Essay in Method." In *Millennium: Journal of International Studies*. Vol. 12, No. 12, 1983.

Crush, Jonathan. *Power of Development*. Routledge, London and New York, 1995.

Cumberland, Charles C. *The Meaning of the Mexican Revolution*. Boston: D.C. Heath and Company, 1967.

Cymet, David. *From Ejido to Metropolis, Another Path. An Evaluation on Ejido Property Rights and Informal Land Development in Mexico*

City. American University Studies, Series 21, Regional Studies, Vol. 6. New York: Peter Lang, 1992.

Darwent, D.F. "Growth Poles and Growth Centers in Regional Planning: A Review." In John Friedmann and William Alonso, eds. *Regional Policy: Readings in Theory and Applications.* Cambridge, Mass.: MIT Press, 1975.

Davis, Kevin and Michael J. Trebilcock. "What Role Do Legal Institutions Play in Development?" Draft prepared for the International Monetary Fund's Conference on Second Generation Reforms, November 8–9, 1999. Np: October 20, 1999.

de Gortari, Ludka. "Comunidad Como Forma de Tenencia de la Tierra" in *Revista de la Procuraduría Agraria* No. 8, July-September. México: Procuraduría Agraria, 1997.

De Janvry, Alain, Gustavo Gordillo and Elisabeth Sadoulet. *Mexico's Second Agrarian Reform. Household and Community Responses.* San Diego: U.S.-Mexican Studies, University of California, 1997.

de Mendizábal, Othón. "Origen de nuestras clases medias." In *Ensayos sobre las clases Sociales en México.* México: Nuestro Tiempo, 1970, 14. Quoted in Roger Bartra. *Estructura Agraria y Clases Sociales en México*, 119. México, D.F.: Ediciones ERA and Instituto de Investigaciones Sociales de la Universidad Nacional Autónoma de México, 1974, repr., 1991.

De Walt, Billie and Martha Reese with Arthur Murphy. *The End of Agrarian Reform: Past Lessons, Future Prospects.* Transformation of Rural Mexico, Number 3, Ejido Research Project. San Diego: Center for U.S.-Mexico Studies, 1994.

Deininger, Klaus and Hans Binswanger. *The Evolution of the World Bank's Land Policy*, Washington, D.C.: World Bank, 1998.

Denzin, Norman and Yvonna Lincoln. *Handbook of Qualitative Research.* Thousand Oaks, CA: Sage, 1994.

DeWalt, Billie R. and Martha W. Rees with Arthur Murphy. *The End of Agrarian Reform in Mexico. Past Lessons, Future Prospects.* Transformation of Rural Mexico, Number 3, Ejido Research Project. San Diego: Center for U.S.-Mexican Studies, University of California, 1994.

Doing Business in Mexico, New York: Price Waterhouse, 1992

Eckstein, Salomón. *El ejido colectivo en México.* México: Fondo de Cultura Económica, 1984.

El Sur, Acapulco, Guerrero, April-May 1994.

Embassy of Mexico in Washington. *Background Information: The Legal Proposal for Mexico's Agricultural Reform.* Washington, D.C.: Embassy of Mexico, Office for Press and Public Affairs, November 1991.

Erfani, Julie A. *The Paradox of the Mexican State: Rereading Sovereignty from Independence to NAFTA.* Boulder, Colorado: Lynne Rienner Publishers, 1995.

Escobar, Arturo. *Encountering Development: The Making and Unmaking of the Third World*. Princeton, NJ: Princeton University Press, 1995.

Estados Unidos Mexicanos, Poder Ejecutivo Federal. *Plan Nacional de Desarrollo 1989–1994*. México: Secretaría de Programación y Presupuesto, 1989.

———. *Plan Nacional de Desarrollo, 1983–1989*. México, D.F.: Secretaría de Programación y Presupuesto, 1983.

Falcón Gutiérrez, José Tomás. *Minería, Comercio y Poder: Los Criollos en el Desarrollo Económico y Político del Guanajuato de las Postrimerías del Siglo XVIII*. Guanajuato, México: La Rana, 1998.

Fanelly, José Maria, Roberto Frenkel and Guillermo Rozenwurcel. "Growth and Structural Reform in Latin America: Where We Stand" in Alvaro Antonio Zini. *The Market and the State in Economic Development in the 1990s*. North Holland, B.V.: Elsevier Science Publisher, 1992.

Flores, Oscar. *México Minero 1796–1950: Empresarios, Trabajadores e Industria*. México: Editorial Font, 1994.

Fox, Jonathan and Gustavo Gordillo. "Between State and Market: The Campesinos' Quest for Autonomy." In Wayne Cornelius, Judith Gentleman and Peter H. Smith, eds. *Mexico's Alternative Political Futures*. Monograph Series 30. La Jolla: Center for U.S.-Mexican Studies, University of California-San Diego, 1989.

Fox, Jonathan. "Political Change in Mexico's New Peasant Economy." In María Lorena Cook, Kevin J. Middlebrook, and Juan Molinar Horcasitas, eds. *The Politics of Economic Restructuring in Mexico: State-Society Relations and Regime Change in Mexico*. San Diego: Center for U.S.-Mexican Studies, University of California, 1994.

Friedman, John and Williams Alonso, eds. *Regional Development Planning: A Reader*. Cambridge, Mass.: MIT Press, 1964.

Garza, Gustavo. *Desconcentración, Tecnología y Localización Industrial en México: Los Parques y Ciudades Industriales, 1953–1988*. México, D.F.: El Colegio de México, 1992.

Gates, Marilyn. "The Debt Crisis and Economic Restructuring: Prospects for Mexican Agriculture." In Gerardo Otero. *Neoliberalism Revisited:Economic Restructuring and Mexico's Political Future*. Simon Fraser University and Westview Press, 1996.

Gilbert, Alan. *The Latin American City*. London and New York: Monthly Review Press, 1994.

Gilly, Adolfo. *Interpretaciones de la Revolución Mexicana*. First repr. México: Universidad Nacional Autónoma de México [UNAM] and Nueva Imagen, 1985.

Gilpin, Robert. *The Political Economy of International Relations*. Princeton, NJ: Princeton University Press, 1988.

Goldring, Luin. "The Changing Configuration of Property Rights Under Ejido Reform." In Laura Randall ed. *Reforming Mexico's Agrarian Reform*, Armonk, New York: M. E. Sharpe, 1996.

Gomezjara, Francisco A. *Bonapartismo y Lucha Campesina en la Costa Grande de Guerrero*. México, D.F.: Editorial Posada, 1979.

Griffit-Jones, Stephany and Osvaldo Sunkel. *Debt and Development Crisis in Latin America: The End of an Illusion*. Oxford, Clarendon: Oxford University Press, 1989.

Grosse, Robert, ed. *Private Sector Solutions to the Latin American Debt Problem*. New Brunswick, U.S.A and London, U.K.: North-South Center, University of Miami and Transactions Publishers, 1992.

Gurría T., José Ángel. "La Política de Deuda Externa de México, 1982–1990." In Carlos Bazdrech, Nisso Bucay, Soledad Loaeza and Nora Lustig. *México: Auge, Crisis y Ajuste*. Macroeconomía y Deuda Externa, 1982–1989, vol. 2. México, D.F.: Fondo de Cultura Económica, 1992.

Hardoy, Jorge. *Urban Planning in Pre-Columbian America*. Planning and Cities, George R. Collins, ed. George Braziller: New York, 1998.

Hiernaux, Daniel. "De Frente a la Modernización: Hacia una Nueva Geografía de México." In Mario Bassols, ed. *Campo y Ciudad en una Era de Transición. Problemas, Tendencias y Desafíos*.México, D.F.: Universidad Autónoma Metropolitana-Iztapalapa, 1994.

Ibarra Mendívil, Jorge Luis. *Propiedad Agraria y Sistema Político en México*. México: El Colegio de Sonora and Miguel Ángel Porrúa, 1989.

Instituto Nacional de Estadística, Geografía e Informática (INEGI). *San Luis Río Colorado, Estado de Sonora. Cuaderno Estadístico Municipal. Edición 1994*. Aguascalientes, México: INEGI, 1995.

———. *Acapulco de Juárez, Estado de Guerrero: Cuaderno Municipal, Edición 1993*. Aguascalientes, México: INEGI, 1994.

———. *Juchitán de Zaragoza, Estado de Oaxaca, Cuaderno Estadístico Municipal, Edición 1995*. Aguascalientes, México: INEGI, 1995.

International Monetary Fund. *Letters of Intent*. htpp://www.imf.org.

Jacobs, Jane M. *Edge of Empire: Postcolonialism and the City*. London and New York: Routledge, 1996.

Johns, Michael. *The City of Mexico in the Age of Díaz*. Austin: University of Texas Press, 1997.

Jones, Gareth A. and Peter M. Ward. "Deregulating the Ejido: The Impact on Urban Development in Mexico." In Wayne Cornelius and David Myhre, eds. *The Transformation of Rural Mexico: Reforming the Ejido Sector*. San Diego: Center for U.S.-Mexican Studies, University of California, 1998.

Katz, Friedrich. "The Agrarian Policies and Ideas of the Revolutionary Mexican Factions Led by Emiliano Zapata, Pancho Villa, and Venustiano Carranza." In Laura Randal, ed. *Reforming Mexico's Agrarian Reform*. Armonk, New York: M. E. Sharpe, 1996.

King, Anthony. *Urbanism, Colonialism and the World Economy; Cultural and Spatial Foundations of the World Urban System*. London and New York: Routledge, 1990.

Krugman, Paul. "Economies of Scale, Imperfect Competition, and Trade: An Exposition," Unpublished paper, 1981, 2. Cited by Gilpin, Robert. *The Political Economy of International Relations*. Princeton, NJ: Princeton University Press, 1988.

——. *Strategic Trade Policy and The New International Economics*. Cambridge: MIT Press, 1986. Quoted by Gilpin, Robert. *The Political Economy of International Relations*. Princeton, NJ: Princeton University Press, 1988.

Lasaga, Manuel. "The Road Back to the Marketplace: How Banks Have Managed the LDC Debt-Restructuring Process." In Grosse, Robert, ed. *Private Sector Solutions to the Latin American Debt Problem*. New Brunswick, U.S.A and London, U.K.: North-South Center, University of Miami and Transactions Publishers, 1992.

Lefevre, Henri. *The Production of Space*. Translated by Donald Nicholson-Smith. Oxford, UK and Cambridge, MA: Basil Blackwell, 1991.

Lenin, Vladimir Ilich. *Imperialism: The Highest Stage of Capitalism*. New York: International Publishers, 1939.

León Portilla, Miguel and Ángel María Garibay. *La Visión de los Vencidos: Relaciones Indígenas de la Conquista*. México, D.F.: Universidad Nacional Autónoma de México, 1987.

Ley General de Sociedades Mercantiles; Ley de Sociedades Cooperativas. México, D.F.: Editorial Pac, 1995.

López Tamayo, Nicolás. "La Urbanización de los Ejidos en la Ciudad de México." Paper presented in the Conference *The Urbanization of the Ejido: The Impact of the Reform to Article 27 Upon Real Estate Development and Land Regulation Policies*. Austin: University of Texas, photocopy, February 1994.

López Velarde Vega, Oscar and Catalina Rodríguez Rivera. "Urbanización del Ejido: El Impacto de la Reforma del Artículo 27 Constitucional. Planeación y Respuestas Institucionales a los Procesos de Desarrollo de Tierra Ejidal." Paper presented in the Conference *The Urbanization of the Ejido: The Impact of the Reform to Article 27 Upon Real Estate Development and Land Regulation Policies*. Austin: University of Texas, photocopy, February 1994.

Lorey, David E. *The U.S.-Mexican Border in the Twentieth Century*. Wilmington: Scholarly Resources, 1999.

Macías, Ruth and José Luis Zaragoza. *El Desarrollo Agrario de México y su Marco Jurídico*. México: CNIA, 1980, 647–648. Quoted in Jorge Luis Ibarra Mendívil. *Propiedad Agraria y Sistema Político en México*. México: El Colegio de Sonora and Miguel Ángel Porrúa, 1989, 239–240.

Melé, Patrice. "Solicitamos Se Nos Regule y No Se Nos Desaloje: Reservas Territoriales, Expropiaciones de Tierra Ejidales y Planeación Urbana." Paper presented in the Conference *The Urbanization of the Ejido: The Impact of the Reform to Article 27 Upon Real Estate*

Development and Land Regulation Policies. Austin: University of Texas, photocopy, February 1994.

México a Través de los Informes Presidenciales (Política Agraria), quoted in Alejandro Carrillo Castro, *La Reforma Administrativa en México: Evolución de la Reforma Administativa en México (1971–1979).* México, D.F.: Miguel Ángel Porrúa, 1980, quote in Erfani, Julie A. *The Paradox of the Mexican State: Rereading Sovereignty from Independence to NAFTA.* Boulder, Colorado: Lynne Rienner Publishers, 1995.

Middlebrook, Kevin J. *The Paradox of the Mexican Revolution.* Baltimore and London: The Johns Hopkins University Press, 1995.

Myhre, David. "Appropriate Agricultural Credit: A Missing Piece of Agrarian Reform in Mexico." In Laura Randall ed. *Reforming Mexico's Agrarian Reform,* Armonk, New York: M. E. Sharpe, 1996.

Nuijten, Monique. *In the Name of the Land: Organization, Transnationalism, and the Culture of the State in a Mexican Ejido.* The Hague, Netherlands: Cip-Data Koninklijke Bibliotheek, 1998.

Nunnenkamp, Peter. "Economic Policies and Attractiveness for Foreign Capital: The Experience of Highly Indebted Latin American Countries." In Hermann Sautter, ed. *Economic Reforms in Latin America.* Frankfurt am Main: Symposium held in November 1992 at the Georg-August-Universität Göttingen, Vervuert Verlag, 1993.

Olmos, Gil. "Abstencionismo, Seguro Vencedor en Comicios." *La Jornada* (México), 28 September 1996.

Osterhammel, Jürgen. *Colonialism: A Theoretical Overview,* Trans. Shelley L. Frisch. Princeton: Markus Wiener Publishers, 1997.

Otero, Gerardo. *Neoliberalism Revisited: Economic Restructuring and Mexico's Political Future.* Simon Fraser University and Westview Press, 1996.

Palpitaciones Porteñas, Acapulco, Guerrero, México, November 1949, quoted by Gomezjara, *Bonapartismo y Lucha Campesina en la Costa Grande de Guerrero.* México, D.F.: Editorial Posada, 1979.

Peña, Devon. *The Terror of the Machine: Technology, Work, Gender, & Ecology on the U.S.-Mexico Border.* Austin: Center for Mexican American Studies, the University of Texas, 1997.

Pérez Rosales, Laura. *Minería y Sociedad en Taxco Durante el Siglo XVIII.* México, D.F.: Universidad Iberoamericana, 1996.

Perroux, François. "Economic Space: Theory and Applications." In John Friedman and Williams Alonso, eds. *Regional Development Planning: A Reader.* Cambridge, Mass.: MIT Press, 1964, 21–36.

Platteau, Jean-Phillipe. *Reforming Land Rights in Sub-Saharan Africa: Issues of Efficiency and Equity* (UNRISD: United Nations Research Institute for Social Development, Discussion Paper No. 60, March 1995). Quoted by Davis and Trebilcock.

Preuße, Gert. "Mexico and the North American Free Trade Association (NAFTA)." In Hermann Sautter, ed. *Economic Reforms in Latin America*. Frankfurt am Main: Symposium held in November 1992 at the Georg-August-Universität Göttingen, Vervuert Verlag, 1993.

Procuraduría Agraria and Programa de Certificación de Derechos Ejidales y Titulación de Solares Urbanos. *¿Qué es y Cómo Funciona el PROCEDE?* Crónicas del PROCEDE 8. México, D.F.: 1993.

Procuraduría Agraria. *Ejido San Luis, El Pionero del Procede*, photo-copied material provided by the Procuraduría Agraria (Hermosillo, Sonora: México, August, 1994).

———. *¿Qué es el PROCEDE?* México, D.F.: 1993.

———. *Informe Anual de Actividades*. México, D.F.: Procuraduría Agraria, 1998. Also found in Web page of the Procuraduría Agraria at http://www.pa.gob.mx/publica/pa07c001.htm#F01

———. *Nueva Legislación Agraria*. 2nd ed. México: D.F.: Unidad de Comunicación Social de la Procuraduría Agraria, 1993.

———. *San Luis Río Colorado, Sonora*. Crónicas de Procede #3, México: Imprenta Juventud, Procuraduría Agraria, Mayo 1994.

Promotora Turística de Guerrero (PROTUR), photocopied material provided by PROTUR, n/d, n/p.

———. *Acapulco Diamante: El Acapulco del Siglo XXI*, promotional material provided by PROTUR, n/d, n/p.

Puiggross, Adolfo. *La España que Conquistó al Nuevo Mundo*. México, D.F.: Costa Amic, 1983.

Ramamurti, Ravi. "Privatization and the Latin American Debt." In Robert Grosse, ed. *Private Sector Solutions to the Latin American Debt Problem*. New Brunswick, U.S.A and London, U.K.: North-South Center, University of Miami and Transactions Publishers, 1992.

Ramírez, Miguel D. "The Political Economy of Privatization in Mexico." Occasional Paper No. 1, presented at the conference *Mexico Beyond NAFTA*. University of Massachusetts, Amherst, Latin American Consortium of New England, n.d.

Randall, Laura ed. *Reforming Mexico's Agrarian Reform*, Armonk, New York: M. E. Sharpe, 1996.

Rébora Togno, Alberto. "Notas Sobre la Acción Gubernamental en el Combate al Precarismo" Paper presented in the Conference *The Urbanization of the Ejido: The Impact of the Reform to Article 27 Upon Real Estate Development and Land Regulation Policies*. Austin: University of Texas, photocopy, February 1994.

Reveles, José, "Punta Diamante, Contraste de Riqueza y Miseria." *El Financiero*, Saturday, 12 November 1994.

Reyna Aoyama, Leticia ed. *Economía Contra Sociedad: El Itsmo de Tehuantepec 1907–1986*, México: Nueva Imagen, 1994.

Reynolds, Clark W. "¿Una Generación Perdida? ¿Por Qué el Desarrollo Latinoamericano Depende del Crecimiento?" In Carlos Bazdrech,

Nisso Bucay, Soledad Loaeza and Nora Lustig. *México: Auge, Crisis y Ajuste*. Macroeconomía y Deuda Externa, 1982–1989, vol. 2. México, D.F.: Fondo de Cultura Económica, 1992.

Ricardo, David, "On Foreign Trade" in *The Principles of Political Economy and Taxation*, First Published in 1821, Third Edition, Amherst and New York: Prometheus Books, 1996.

Roberts, Bryan. "The Place of Regions in Mexico" in Van Young, Eric, ed., *Mexico's Regions: Comparative History and Development*. San Diego: Center for U.S.-Mexican Studies, University of California, 1992.

———. *Cities of Peasants*. London: Edward Arnold, 1978.

———. *The Making of Citizens. Cities of Peasants Revisited*. London: Arnold, 1995.

Robles Berlanga, Héctor Manuel. "Tipología de los Sujetos Agrarios Procede." *Revista de la Procuraduría Agraria* 4, July-September 1996. Also found in web page of the Procuraduría Agraria at http://200.38/178/160/publica/pa/pa070403.htm

Robles, Rosario and Julio Moguel. "Agricultura y Proyecto Neoliberal." *El Cotidiano* (México), April 1990, 3. Quoted in Tom Barry, ed. Mexico: A Country Guide. The Inter-Hemispheric Education Resource Center. Albuquerque, New Mexico, 1992.

Ross Schneider, Ben. "The Politics of Privatization in Brazil and Mexico: Variations on a Statist Theme." Conference Paper No. 23. New York: A National Resource Center for Latin American and Caribbean Studies, The Columbia University/New York University Consortium, 1990.

Rossi, Peter H., Howard E. Freeman and Mark W. Lipsey. *Evaluation, A Systemic Approach*. Newbury Park, CA: Sage Publications, 149.

Rudolph, James. *Mexico: A Country Study*. Foreign Area Studies. Washington, D.C.: The American University, 1985.

Ruiz Arrazola, Victor. "El Senado Pidió Preferir el Capital Mexicano al Transnacional: Ruiz Sacristán," in *La Jornada*, July 23, 1996.

Said, Edward W. *Orientalism*. New York: Vintage Books, 1979.

Sánchez López, Héctor Manuel. *Estudio Monográfico de la Comunidad de Asunción Ixtaltepec, Juchitán, Oaxaca*. Photocopy, March 1975.

Sautter, Hermann, ed. *Economic Reforms in Latin America*. Frankfurt am Main: Symposium held in November 1992 at the Georg-August-Universität Göttingen, Vervuert Verlag, 1993.

Schulze-Gaevernits. *Britischer Imperialismus*. Quoted by Vladimir Ilich Lenin, *Imperialism The Highest Stage of Capitalism*. New York: International Publishers, 1939.

Schwandt, Thomas A. *Qualitative Inquiry: A Dictionary of Terms*. Thousand Oaks, CA: Sage, 1997.

Scott, Ian. *Urban and Spatial Development in Mexico*. Baltimore: World Bank, Johns Hopkins University Press, 1982.

Secretaría de Agricultura, Ganadería, Desarrollo Rural, Pesca y Alimentación. "PROCAMPO," *Programas de Apoyo.* http://www.sagarpa.gob.mx/sagar2.htm.

Secretaría de Desarrollo Social (SEDESOL), Colegio de Arquitectos and Instituto de Investigaciones Económicas. *México 2020. Un Enfoque Territorial de Desarrollo; Vertiente Urbana. Síntesis Ejecutiva.* México, D.F.: SEDUE, 2000.

———. *Programa de 100 Ciudades 1990–2000*, (January 1993), n.p.

Secretaría de la Reforma Agraria. *Comunicado de Prensa* No. SRA/006. Unidad de Comunicación Social, Boletín de Prensa, January 29, 2001.

———. *Comunicado de Prensa* No. SRA/007. Unidad de Comunicación Social, Boletín de Prensa, January 31, 2001.

Silva Herzog, Jesús. *El Agrarismo Mexicano y la Reforma Agraria.* México: Fondo de Cultura Económica, 1980, 70–73 and 78–82. Quoted in Jorge Luis Ibarra Mendívil. *Propiedad Agraria y Sistema Político en México.* México: El Colegio de Sonora and Miguel Ángel Porrúa, 1989, 91.

Simbieda, William. "Social Land and Urban Needs: Ejido Transformation at the Periphery." Paper presented in the conference *The Urbanization of the Ejido: The Impact of the Reform to Article 27 Upon Real Estate Development and Land Regulation Policies.* Austin: University of Texas, photocopy, February 1994.

Smith, Wesley R. "Salinas Prepares Mexican Agriculture for Free Trade." *Backgrounder* #914. Washington, D.C.: The Heritage Foundation, October 1992.

Soberanes Reyes, José Luis. *La Reforma Urbana. Una Visión de la Modernización de México.* México, D.F.: Fondo de Cultura Económica, 1993.

Soja, Edward W., *Thirdspace: Journey to Los Angeles and Other Real-And-Imagined Places.* Cambridge, MA and Oxford, UK: Blackwell Publishers, 1996.

Stake, R.E. *The Art of Case Study.* Thousand Oaks, CA: Sage, 1995.

Stephen, Lynn. *Viva Zapata!: Generation, Gender, and Historical Consciousness in the Reception of Ejido Reform in Oaxaca.* The Transformation of Rural Mexico No. 6, Ejido Research Project. San Diego: Center for U.S.-Mexican Studies, University of California, 1994.

Tannenbaum, Frank. *Mexico: The Struggle for Peace and Bread.* New York: Borzoi Books published by Alfred A. Knopf, 1951.

Thiesenhusen, William C. "Mexican Land Reform, 1934–91: Success or Failure?" In Laura Randall ed. *Reforming Mexico's Agrarian Reform.* New York: M. E. Sharpe, 1996.

Vázquez García, Verónica ed. *Género, Sustentabilidad y Cambio Social en el México Rural.* Estado de México, México: Colegio de Postgraduados, 1999.

Vázquez Ruiz, Miguel Ángel ed. *Sonora hacia el 2000: Tendencias y Desafíos.* Hermosillo, Sonora: Ed. Sino, 1993.

Vickers, John and George Yarrow. *Privatization: An Economic Analysis.* Mass: MIT Press Series on the Regulation of Economic Activity #18, 1988. Cited by Ravi Ramamurti. "Privatization and the Latin American Debt." In Robert Grosse, ed. *Private Sector Solutions to the Latin American Debt Problem.* New Brunswick, U.S.A and London, U.K.: North-South Center, University of Miami and Transactions Publishers, 1992.

Vidaurri, Miguel Ángel. "Principales Resultados del Censo de Órganos de Representación de Ejidos and Comunidades." Revista de la Procuraduría Agraria 10, May-December 1998. Also found in web page of the Procuraduría Agraria at http://200.38/178/160/publica/pa071007.htm

Ward, Peter. *Mexico City: The Production and Reproduction of an Urban Environment.* John Wiley and Sons, 1998.

Watchel, Howard. M. *The Money Mandarins: The Making of a Supranational Economic Order.* New York: Sharpe, Inc., 1990.

Waters, Malcolm. *Globalization.* London and New York: Routledge, 1995.

Weintraub, Sidney. ed. *Integrating the Americas: Shaping Future Trade Policy.* New Brunswick: Transaction Publishers and North/South Center, University of Miami, 1994.

Williams, John. *The Progress of Policy Reform in Latin America.* Institute for International Economics, Washington, D.C.: 1990.

Wolf, Eric W. "El Bajío en el Siglo XVIII: Un Análisis de Integración Cultural." In David Barkin, ed. *Los Beneficiarios del Desarrollo Regional.* México, D.F.: SepSetentas 52, Edimex, 63–95.

Wright, Harry. *Foreign Enterprise in Mexico: Laws and Policies.* Chapel Hill: University of North Carolina Press, 1971.

Yin, Robert. *Case Study Research: Design and Methods.* Second edition. Thousand Oaks, CA: Sage, 1994.

Zepeda Lecuona, Guillermo R. "Cuatro Años de Procede: Avances y Desafíos en la Definición de Derechos Agrarios en México." *Revista de la Procuraduría Agraria* No. 9, October 1997–April 1998.

Zini, Alvaro Antonio. *The Market and the State in Economic Development in the 1990s.* North Holland, B.V.: Elsevier Science Publisher, 1992.

Index